Environmental Footprints and Eco-design of Products and Processes

Series editor

Subramanian Senthilkannan Muthu, SGS Hong Kong Limited, Hong Kong, Hong Kong SAR

More information about this series at http://www.springer.com/series/13340

Subramanian Senthilkannan Muthu
Monica Mahesh Savalani
Editors

Handbook of Sustainability in Additive Manufacturing

Volume 2

 Springer

Editors
Subramanian Senthilkannan Muthu
Environmental Services Manager-Asia
SGS Hong Kong Limited
Hong Kong
Hong Kong

Monica Mahesh Savalani
Department of Industrial and Systems
  Engineering
The Hong Kong Polytechnic University
Hong Kong
Hong Kong

ISSN 2345-7651   ISSN 2345-766X (electronic)
Environmental Footprints and Eco-design of Products and Processes
ISBN 978-981-10-0604-3   ISBN 978-981-10-0606-7 (eBook)
DOI 10.1007/978-981-10-0606-7

Library of Congress Control Number: 2016930176

Printed on acid-free paper

This Springer imprint is published by SpringerNature
The registered company is Springer Science+Business Media Singapore Pte Ltd.

Contents

Introduction

This *Handbook of Sustainability in Additive Manufacturing* is planned to be published in two volumes and the current one is the second volume which carries five very informative and well-written chapters. Dealing with the core concepts of the subject, these five topics are written by experts in the field.

Chapter "Environmental Impact Assessment Studies in Additive Manufacturing", authored by Dr. Olivier Kerbrat, Dr. Florent Le Bourhis, Pr. Pascal Mognol, and Pr. Jean-Yves Hascoët revolves around characterizing the environmental impact of additive manufacturing processes. In addition, the authors study a new methodology which is used to evaluate the environmental impact of a part from its CAD model. An industrial example is also studied and it shows that material consumption has an important impact and has to be taken into consideration for a complete environmental impact assessment.

Chapter "Sustainability Based on Biomimetic Design Models", written by Henrique A. Almedia and Mario S. Correia focuses on works which aim to evaluate the ecological impact of the support production methodologies in order to deliver awareness to the users of extrusion-based systems for a lower environmental impact assessment. The extra production time involved in the production of the support structures and the support structure removal is evaluated.

Chapter "Energy Efficiency of Metallic Powder Bed Additive Manufacturing Processes", authored by Konstantinos Salonitis discusses the sustainability issues in relation to the processing of metallic powders with additive manufacturing processes. In addition, the sustainability at various levels was studied to show the variations. Furthermore, the basics for modelling the finite element process were discussed in relation to the energy efficiency and the optimization of the process. In general, during the processing of metallic powders, the preheating of the raw material is the key energy consumer; alternative methods of preheating were discussed and compared.

Chapter "Carbon Footprint Assessment of Additive Manufacturing: Flat and Curved Layer-by-Layer Approaches", written by Subramanian Senthilkannan Muthu and Savalani Monica Mahesh describes and discusses the carbon footprint assessment of the fused deposition modelling process and its adapted technique, namely the curved fused deposition modelling process. Preliminary investigations have indicated that curved FDM parts are stronger than those made using the FDM process. However, the initial carbon footprint does indicate that the material usage and processing time of the curved FDM process are higher than those of the FDM process. The authors have recommended a fuller investigation of the sustainability issues of these two processes before in-depth conclusions can be drawn.

Chapter "Sustainable Frugal Design Using 3D Printing", authored by Ian Gibson and Abhijeet Shukla describes and discusses sustainable solutions which can be achieved through the fusion of 3D printing with frugal approaches in design and engineering. 3D printing technologies have forced many to rethink how we design and turn this group of technologies and processes into a new business. The chapter explores the product design and development process in relation to frugal and sustainable concepts and how 3D printing and related technologies influence them. This chapter highlights how frugal designs enable designing for low-cost solutions to many important problems and how this could also assist with the development of low-cost and accessible machines which use recycled materials.

The editors would also like to express their gratitude to all the contributing authors.

Subramanian Senthilkannan Muthu
Monica Mahesh Savalani

Energy Efficiency of Metallic Powder Bed Additive Manufacturing Processes

Konstantinos Salonitis

Abstract Metallic powder bed additive manufacturing processes have evolved a lot over the last few years. A number of alternative processes have been developed and are classified in the present chapter. In order for these processes to deliver metallic parts, a large amount of energy is delivered to the powder that is used as a raw material. The implications to the sustainability are discussed and sustainability key performance indicators are presented. This chapter presents a comprehensive review of the relevant literature of the studies presented on the energy efficiency of metallic powder bed additive manufacturing processes and the key challenges for improving it. Furthermore, modelling of the process with finite element simulation is discussed for the estimation of the energy efficiency and the optimisation of the process under this prism. Because the preheating of the raw material is the key energy consumer, the alternative methods are discussed and compared.

Keywords Additive manufacturing · Energy efficiency, modelling

1 Introduction

Additive manufacturing (AM) processes were developed in the mid-1980s and since then a number of processes have been presented. The key advantage of such processes is the freedom that they provide to the designers for coming up with product designs that cannot be manufactured conventionally. Conventional manufacturing processes, such as machining, pose limitations on the component geometries that can be produced. These limitations often result in structures that are inefficient, as many areas of a component have excess material that cannot be removed physically or in a cost-effective way through conventional methods. AM allows components to be manufactured in a bottom-up approach with laying

K. Salonitis (✉)
Manufacturing Department, Cranfield University, College Road, B50,
Cranfield MK43 0AL, UK
e-mail: k.salonitis@cranfield.ac.uk

© Springer Science+Business Media Singapore 2016
S.S. Muthu and M.M. Savalani (eds.), *Handbook of Sustainability in Additive Manufacturing*, Environmental Footprints and Eco-design of Products and Processes, DOI 10.1007/978-981-10-0606-7_1

1

material only where it is required. The main technological breakthrough that allowed the development of both desktop processes as well as AM, was the evolution of computer and information technology, and especially the various computer-assisted technologies in the manufacturing area, such as computer-aided design (CAD), computer-aided manufacturing (CAM), and computer-aided engineering (CAE). One of the key advantages of such processes thus is the fabrication of components and even complete assemblies directly derived from CAD models without the need for process planning in advance of manufacturing. Various methods that allow the 'building' of three-dimensional objects in sequence by adding layers over each other have been developed [1].

AM technology has a relatively short history of about 25 years and it has grown largely since its invention. According to the *Wohlers Report* [2], the AM projected value for 2015 is $4 bn, and will reach $6 bn in 2017 and almost $11 bn in 2021. However, although it is becoming more and more mature, and often claimed as the 'next industrial revolution' there are still a number of challenges for successful commercialisation. AM technology challenges are related to the materials, the available CAD software, the data management, the sustainability, the affordability, the process speed, the process reliability, the intellectual property, and the standards to name a few [3]. The design for AM has been also identified as a key challenge, highlighting that for exploiting the capabilities that AM processes offer, the designers have to adapt their approach to the AM technology, not replicating the existing methods and philosophies established for conventional processes.

Additive manufacturing is the process of fabricating an artefact (geometrically defined product), which is derived from a model, initially generated using a three-dimensional computer-aided design (3D CAD) system, directly without the need for process planning. Rapid prototyping belongs directly to additive manufacturing, but differentiates itself through the intended means of rapid prototyping, which is described as 'processes for rapidly creating a system or part representation (like a prototype) before final release or commercialization'. Additive manufacturing processes can be classified in three different categories depending on the status of the material used to create the artefact during the process, for example, powder based, liquid based, and solid based as depicted in Fig. 1. A large number of different AM processes have been developed in the short history of AM; few of them though survived over time. Common materials are aluminium, steel alloys, precious metals, plastics used in a powder form, and paper; but wood, wax, paper, clay, concrete, sugar, and chocolate are possible to be used as filament. The most widespread AM processes include:

- *Selective laser sintering (SLS)* is a sintering process where a surface of a powder bed is fused with a high-power laser. Thus, a geometrically defined product is built up layer by layer and the nonmelted powder is cleaned away and can be recycled. SLS is suitable for highly complex parts that contain undercuts because the powder is supporting the structure.
- *Laser engineering net shaping (LENS)* is a process where a geometrically defined product is built by injecting metal powder into a molten pool. The pool

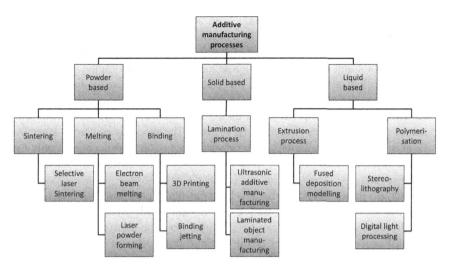

Fig. 1 Additive manufacturing processes classification (updated from Kruth et al. [1])

is created by a focused, high-powered laser beam and the part is fabricated by depositing cross-sections of material layer by layer. LENS can also be used to repair parts but still needs post-production treatments.

- *Electronic beam melting (EBM)* is a process that takes place in a vacuum, where a high-power electron beam is used to melt the surface of a metal powder bed in layers. As the molten surface cures, the nonliquefied powder is cleared away, thus generating the geometrically defined product. Compared to the SLS process, EBM fabricates fully dense metal components with better material properties.
- *3D printing* is a process of building a model by spraying a binder on the cross-section parts of a powdered material. An elevator lowers the printing bed and a new layer of powder is spread. Finally the part is cured and the unbounded powder is removed. 3DP is a fast process with low material costs though other costs are much higher and parts have a rough surface finish.
- *Prometal* is a process where a geometrically defined product is built by dripping a binder or resin on a thin layer of powdered metal, glass, or sand. An elevator lowers the printing bed and a new layer of material is spread. Finally, the part is cured and the unbounded powder is removed. The Prometal process can fabricate complex sand casting moulds.
- *Laminated object manufacturing (LOM)* is a process where stacked cross-sections of adhesive-coated sheet material are used to build a geometrically defined product. The sheet material (paper, plastics, and metals) is adhered to the subjacent sheet and then cut by a laser beam. The nonpart areas get removed and the next sheet is stacked for the following cross-section.

- *Fused deposition modelling (FDM)* is an additive manufacturing process that builds a model, layer by layer, by extruding material from a movable FDM nozzle. The raw material is unrolled from a coil, heated up slightly above its melting point and then deposited by an extrusion head. Common 3D printers, commercially available, fall under this category even though the name wrongfully implies that they fall under the category of 3DP.
- *Stereolithography* is a process that uses a liquid photosensitive monomer resin that is solidified by an ultraviolet (UV) laser. The model is produced layer by layer in a bath of resin in which the upper surface of the part is covered only by a thin layer of resin. The UV laser then solidifies the surface by tracing the cross-section.
- *Polyjet* is a printing process that jets layers of liquid photopolymer and support material on the printing tray and UV bulbs cure each layer after depositing. Finally the support material can be separated from the part (i.e., by water jet). For this process no additional curing or finishing treatments are needed.

A large number of different AM processes for fabricating metallic parts, utilising different combinations of stock material form, material delivery, and heat source have been developed over the years. In the powder-based processes, metallic powder is spread on the bed before being scanned by the beam or being fed directly to the heat source affected region [4]. The powder bed processes can be further classified based on whether the stock material gets fully melted, partially melted, or a polymer binder is used for consolidation [5–7]. Over the previous decades a significant number of different processes, covering all of the aforementioned categories, have been developed. A classification schematic listing all those processes is provided in Fig. 2, where the types of processes discussed in this chapter are highlighted with orange colour. Despite the numerous powder bed processes, the most commercialised among them are the selective laser melting, electron beam melting, and selective laser sintering.

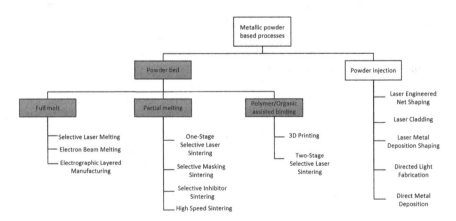

Fig. 2 Classification of metallic powder additive manufacturing processes

1.1 Challenges of the Powder Bed AM Processes

AM process mechanisms present a high level of complexity; the physical mechanisms are still under investigation and a lot of research is undertaken for better comprehension. Therefore, in many cases the programming of the process is based on trial and error and not on simulation. As a result, the parts produced using AM processes often do not meet the mechanical performance requirements set by manufacturers and/or exhibit difficulty in predicting the distortion of the final part. This, along with the long lead times, further adds to the challenges identified [3] already. Academia, however, is investing a lot of effort and resources for overcoming these problems that will eventually allow the adoption of AM when it comes to the production of metallic parts. By applying the '5 whys' root cause analysis, the key challenges can be identified. As an example, the poor mechanical performance of the final product can be attributed to the porosity and the presence of thermally induced residual stresses and distortion [8]. Kruth et al. [9] proved that the porosity of the final part is a result of the process instabilities during the material consolidation. Such instabilities, referred to as 'balling', prevent a full surface contact between melted material particles as shown by Gu et al. [10], leading to parts with lower density than the respective solid material and therefore decreased mechanical strength. The aforementioned residual stresses and geometrical distortion are due to the thermomechanical nature of the process itself and the thermal gradients created when building the part. Schoinochoritis et al. [11] summarised these challenges in a root cause diagram as can be seen in Fig. 3. It is interesting to mention that on the 'environment' branch the identified challenges include the 'low energy efficiency' and the 'decreases energy consumption for the life cycle of the product'. These two challenges are covered in detail in the present chapter and the possibility of using simulation and modelling for increasing their performance is assessed.

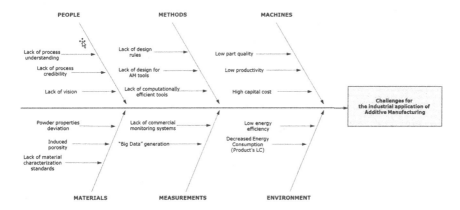

Fig. 3 Challenges in additive manufacturing industrial application (adapted from [11])

1.2 Structure of the Chapter

The structure of this chapter is as follows. Section 2 discusses the issues of sustainability and energy efficiency initially for the manufacturing processes in general and then focusing on AM and specifically on powder bed AM-based methods. The advantages of AM processes with regard to sustainability are outlined, the fact that AM can further enhance the circular economy paradigm is highlighted, and the energy efficiency of such processes is discussed. Section 3 discusses the modelling methods for the simulation of the AM processes and how these can be used for predicting the energy efficiency of the process. Details about mathematical modelling, meshing, simulation techniques, heat source models, and temperature-dependent material properties implementation into the model are provided. Section 4 focuses on the energy efficiency of AM processes, discussing the need for a structured way of auditing the energy consumption of such processes, the key energy-consuming components of AM machines, and ways of optimising their efficiency. Modelling for predicting energy efficiency is also included in this section. Finally this chapter is summarised and a number of concluding remarks are given in Sect. 5.

2 Sustainability and Energy Efficiency of Manufacturing Processes

Manufacturing processes in general are defined as the processes that transform materials and information into goods for the satisfaction of human needs. However, turning raw materials into consumer products is also a major source of environmental pollution. This environmental pollution can be the direct outcome of the manufacturing process, or indirectly through the use of energy for running these processes. Manufacturing waste involves a very diverse group of substances, and depends on the technology used, the nature of the raw material processed, and the quantity that is discarded at the end of the chain. The large use of energy for industrial operations in Europe (32 % of the whole consumed energy) is responsible for significant CO_2 emissions and thus climate change [12]. AM processes can potentially reduce the waste for the manufacturing of specific products and/or under specific circumstances as is discussed in the present section.

2.1 Sustainability and Additive Manufacturing

From a societal perspective, one of the most common definitions of sustainability and sustainable development was provided by the Brundtland [13] Commission: 'Sustainable development is development that meets the needs of the present without compromising the ability of future generations to meet their own needs'. Generally, sustainability has three common dimensions; social, economic, and environmental. Where, traditionally, economic aspects dominate decision making, with respect to sustainability, these dimensions should be considered simultaneously and equally.

When assessing sustainability of an AM process (and in fact for any manufacturing process), the entire life cycle of the product to be manufactured needs to be considered for concluding with confidence the impact to the sustainability. The actual manufacturing process impact is only one of the many environmental impacts associated with the product life cycle. Mani et al. [14] dealt with the sustainability characterisation for AM processes and identified a number of advantages of AM that are relevant to sustainable manufacturing, listed hereafter:

- The nature of the AM processes (building layer by layer and depositing material only where needed) results in less waste when compared to mass production processes such as forming and machining (where lots of material is scrapped as chips, for example, in the case of machining).
- AM processes do not require any specialised tooling or fixtures, eliminating thus the manufacturing processes for these and the need for disposal at the end of their life cycle.
- Again due to the nature of the AM processes, they are capable of building functionally lightweight parts, while maintaining strength, which has a significant impact on their life-cycle sustainability impact (e.g., reduced weight of specific aerospace components that result in reduced need for energy to move them around).
- Because the AM processes use raw material only when needed, this reduces the need for stocking large amounts of raw material within the supply chain and transportation.
- AM processes are largely material efficient when compared with traditional machining and casting.
- AM processes have the ability to produce optimised geometries with near-perfect (compared with wrought material) strength-to-weight ratios, which is very important for lightweight structures such as airplanes.
- There is less impact of the part over its life cycle, resulting in a lower carbon footprint, less embodied energy, and a better economic model.
- AM processes have the ability to create on-demand spare parts, reducing or eliminating inventory, or the need for transportation of these spare parts over great distances.

2.2 Circular Economy and Additive Manufacturing

The life of a product is quite variable, ranging from few minutes for the case of metal cans to more than 30 years for the case of airplanes and even centuries for buildings and infrastructure. One of the key concepts nowadays relies in prolonging the life of the products for reducing the demand for new material and the environmental impact of production. Prolonging the life of the products subsequently reduces the demand for new material and reduces the environmental impact of production. Additionally, reusing or even refurbishing the parts before recycling and discarding can further decrease energy demand and environmental impact. Based on this prolonging idea, an initiative from the Ellen MacArthur Foundation [15] with regard to the life cycle of manufactured parts suggests the transition from a 'linear to circular economy'. A circular economy seeks to rebuild capital, whether this is financial, manufactured, human, social, or natural. The basic idea is to replace a linear industrial model with one based around reuse and recycling. This ensures enhanced flows of goods and services. In Fig. 4 the circular economy concept for technical materials is presented [16]. Some of the loops involved in a circular economy are shown, indicating how products are designed to be fixable, refurbishable, and recyclable at the end of their lifetimes, 'an industrial system that is restorative by design'. The most sustainable design is the one that lies in the most internal loop.

Gebler et al. [17] analysed a number of market scenarios with regard to the 3D printing technologies evolution. The life-cycle analysis indicated that the adoption of AM could have significant savings in the production and use phases of a product.

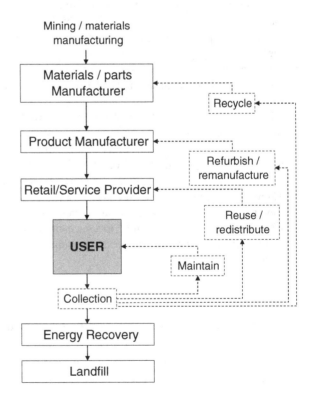

Fig. 4 Circular economy: only technical materials represented

Estimates for the market in 10 years' time (2025) are in the range of $113–370 billion for the production phase and $56–219 billion for the use phase of 3D printed products. As Despeisse and Ford [18] highlighted, the savings in the production phase stem from reduced material inputs and handling, along with shorter supply chains. In the use phase, lightweight components enable energy consumption to be reduced.

Despeisse and Ford [18] focused on the implications of AM technology on resource efficiency and sustainability. Based on a number of industrial examples, four areas were identified where the use of AM could lead to improved performance of the industrial systems with regard to the sustainability of the products, namely product and process design, material input processing, make-to-order products and component manufacturing, and closing the loop as indicated in Fig. 4. With regard to closing the loop, their argument is focused on the fact that repair, maintenance, and remanufacturing phases can be approached better and more efficiently with a make-to-order model for minimising inventory waste as spare parts can be produced locally only when needed, with lower energy intensity processes.

2.3 Sustainability Indicators for AM

There is a need for a number of metrics to be used for assessing the performance of a process with regard to its implications for sustainability. Chen et al. [19] compiled a list of sustainability indicators for AM processes based on three major groups of indicators and indices (that are, however, focused on all the manufacturing activities and the company holistically):

- Dow Jones Sustainability Index to evaluate company sustainability, most related to the economical part of sustainability
- GM Metrics for sustainable manufacturing developed for the car industry
- GRI Reporting Framework developed to evaluate the company and factory

Obviously these indices are not considered practical for assessing the implications of AM in sustainability. Chen et al. [19] collected the sustainability-related implications of AM in comparison to existing manufacturing technologies and systems. Table 1 summarises their findings. It is evident that economic and environmental dimensions are closely related.

With regard to the *economic dimension* of sustainability, AM technology has significant economic potentials. The decentralisation of manufacturing, the direct access to information and design by many individuals who are not necessarily manufacturing or design experts, and their influence on the design of products and services (referred to as democratisation of manufacturing by some scholars) can have a major impact on the way industry is operating and running their businesses. (Just try to answer the question: What if anyone could make almost anything they need, anywhere?) However, a number of challenges (as shown in Fig. 3 as well) still prohibit the substitution of conventional manufacturing (and possibly never

Table 1 Implications of AM on sustainability dimensions

Economic	Environmental	Social
• Higher material utilisation (+) • Simpler, more efficient supply chains with less transportation efforts (+) • Less material and energy losses due to less inventory (+) • Less waste and better waste management through possibility of direct recycling (+) • User oriented manufacturing, less overproduction in stocks (+) • No moulds etc. necessary (+) • Higher specific energy demand (−) • Quality issues are not finally solved, thus risk of bad parts and rework (−)		• Equal possibilities for all participants in markets and societies (+) • Bridge technological, educational, and cultural gaps between developing and developed countries (+) • User-oriented products, more customer satisfaction (+) • Potential benefits on human/worker health (+) • Unclear impact on an employment situation of industry (±)
• Potentially higher profit due to customer-specific solutions (+) • Profitability could be proved in selected cases (±) • Longer manufacturing time (−)	• Ambivalent studies in terms of an environmental impact or eco-efficiency, (±)	

Note Indicates whether the specific aspects of DDM tend to be rather: beneficial (+), unfavourable (−), or whether it cannot clearly be assessed for the general case (±)

Source Adapted by Chen et al. [19]

will) such as quality problems due to the low tolerances; material and processing capabilities also need to be improved. The energy efficiency of the AM processes has a direct impact on the economic dimension, as the energy cost is significant. In a number of studies this has been considered in financial models [20, 21] proving that AM processes can be more cost efficient than conventional manufacturing for small to medium production volumes (when considering the fact that moulds and tools are not required for AM and the waste management is way more cost efficient inasmuch as the waste and scrap are minimal).

The *environmental dimension* of the AM processes is assessed through three major indices, namely the total environmental impact, the carbon dioxide equivalent emissions (related to the greenhouse effect), and the sources of energy. These indices, however, consider the whole life cycle of the product (and thus the raw material as well) and the energy required for the different phases of production, and obviously the metal-based ones usually score worse than the rest of the AM processes. Selective laser sintering, for example, is more than eight times higher than the other processes [22]. Life-cycle analysis (LCA) is usually used for assessing the environmental impact of the manufacturing processes. In one of the few studies where LCA was used for AM processes, it was revealed that AM processes perform better than machining because no chips are produced that require further processing afterwards [23].

It is common practice for manufacturing decision-making systems to address the economic pillar of sustainability and lately increasing effort has been directed towards the environmental pillar via attention to the environmental life-cycle impacts of products and manufacturing processes. However, until recently the pillar associated with the *social dimension* of sustainability has not been well defined and exploited. In contrast to the ideal systems posited by sustainability advocates and manufacturing futurists, conventional manufacturing has negative environmental and social externalities. Manufactured products and capital for production (including infrastructure, machines, tools, and production lines) tend to be created though traditional structures which are inefficient and cause negative impacts. In the literature the issue has not received great attention, from a manufacturing point of view, emphasising mainly legislative and health and safety human issues rather than cultural and ethical decision-making parameters [24–26]. Bell and Morse [27] concluded that attempts to measure social sustainability ends up in an effort attempting to measure the immeasurable. A number of approaches have been presented targeting the quantification of life quality from both objective and subjective perspectives [28]. The wide quantity and variety of social indicators imply the challenge of creating a framework capable of evaluating social implications of manufacturing at a micro level. Similarly, for AM technology a limited number of studies have been presented that deal with the social dimension. Huang et al. [29] presented a study on the impact of AM on the work conditions and the long-term health and social implications for the AM operators concluding that there is a health benefit when compared to the conventional processes such as casting, forging, and machining in terms of avoiding long-term exposure of hazardous noise and oil mist from metal working fluid.

2.4 Energy Efficiency

Defining energy efficiency is critical for assessing the performance of the AM process and AM machine. This thus requires integrating energy efficiency in the existing metrics that are used for measuring the performance of a production system. The performance of a production/manufacturing system is achieved by monitoring four key manufacturing attributes, namely cost, time, quality, and flexibility. Focusing on exploiting the economies of scale that affected the decision-making process for the manufacturing sector as well, the research and development mostly dealt with the cost and time (up till the 1970s) and afterwards on quality and eventually on flexibility. However, optimising for these attributes without considering other aspects (such as energy efficiency and consumption) might result in solutions and technological improvements with higher energy requirements. It is evident that these four attributes do not take into consideration energy or resources efficiency that are key to sustainability. Although it could be claimed that within the cost attribute, the energy cost is hidden as well, however, because the energy cost had been a small fraction of the overall cost, the impact was not evident. Nevertheless, sustainability has evolved to be a key attribute that has to be considered when making manufacturing decisions. Therefore, the manufacturing

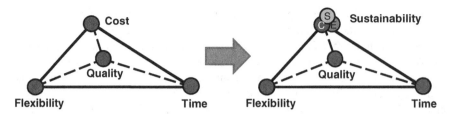

Fig. 5 Manufacturing decision-making attribute evolution (where social (S), environmental (E), and cost/economical (C) are the dimensions of sustainability)

tetrahedron that was proposed by Chryssolouris [30] was extended by Salonitis and Ball [31] to include 'sustainability' as a new driver in manufacturing (Fig. 5).

Salonitis and Ball [31] and Salonitis and Stavropoulos [16] discussed the key performance indicators (KPIs) that should be considered for sustainability. The KPIs for the other manufacturing attributes have been discussed in numerous studies over the years. As an example, the "Cost" attribute incorporates both direct and indirect costs analysed through a number of factors such as equipment and facility costs, material cost, labour, overhead, and the like. The 'Time' attribute is monitored using KPIs such as 'throughput time', 'cycle time', 'lead time', and so on. The 'Quality' attribute is assessed in a number of ways depending on whether we refer to the surface quality, the defects, and the usual KPIs including 'surface roughness' measurements, 'cost of quality', defect rates, yield, and so on. The 'Sustainability' attribute is a relevant new one and is still quite challenging to assess it. Energy efficiency, on the other hand, is discussed based on existing definitions that can be rather misleading [32]. In general, 'energy efficiency' refers to technologies and standard operating procedures that reduce the volume of energy per unit of industrial production. IEA-adopted definitions of energy efficiency are: 'the goal of efforts to reduce the amount of energy required to provide products and services' and 'achieving the same quality and level of some "end use" of energy with a lower level of energy input'. Nevertheless, a number of energy-related KPIs have been introduced and can be categorised into metrics focusing on the energy consumption (such as energy consumed per product, total on-site energy, total energy use, etc.), environmental impact (CO_2 emissions, greenhouse gas emissions, etc.), and financial figures (e.g., energy cost), focusing on the process level, machine tool or production plant, and the like.

3 Modelling of Metallic Powder Bed AM Processes

Schoinochoritis et al. [11] summarised the presented works on the modelling of metallic powder bed AM processes focusing mainly on the finite element analysis ones. In general, the studies presented were focused on predicting the temperature field and subsequently the stress distribution on the additive manufactured part. This thermomechanical analysis can be broken into two separate analyses that can be treated independently for the ease of mathematical processing. However, they need to be coupled as the thermal analysis feeds input to the mechanical one. Both thermal and mechanical analyses were further broken down into three sub-phases: the preprocessing where input is provided to the finite element analysis such as the geometry, the material properties, the boundary conditions (such as the loads and the fixing points), the processing where the mathematical equations are solved, and finally the postprocessing where the results are extracted.

Due to the fact that the physical mechanisms that drive the process are nonlinear, the modelling is performed using finite element analysis. For reaching the solution of the finite element analysis either explicit or implicit methods can be used. Explicit methods can exhibit numerical instability. The advantage of implicit methods is that they are unconditionally stable, hence larger time increments can be used. Small time increments required by explicit methods have an impact on computational cost, in contrast to implicit methods where computational cost is proportional to the size of the finite element model. However, the implicit method may encounter difficulty in converging when the problem is highly nonlinear. In the case of powder-based AM processes, the problem involves nonlinearities due to temperature-dependent material properties and plastic deformations.

For modelling the process in order to estimate the energy efficiency of the process, a very good understanding of the process mechanisms is required. In the following paragraphs these mechanisms are described and modelled.

3.1 Modelling of the Heat Transfer Problem During AM

In additive processes, the part is created by consolidating material in specific regions according to the part geometry. To achieve this, energy is provided by irradiation to the consolidation area. The heat source (usually a laser beam) delivers the energy that results in melting the material and through solidification the geometry is built. The energy is partitioned to the heat used for the melting of the material, and heat that escapes the system to the surroundings by convection and radiation from the free surfaces. For the powder bed AM process, the heat transfer mechanisms acting during the process are illustrated in Fig. 6.

The energy balance equation considering the first law of thermodynamics, as graphically shown in Fig. 6, can be expressed by the following equation,

$$Q_L = Q_{CD} + Q_{CV} + Q_R \tag{1}$$

where Q_L, Q_{CD}, Q_{CV}, Q_R are the heat flux from the heat source, and the heat loss due to conduction, due to convection, and due to radiation, respectively.

In most of the studies, the research is focused on the heat transferred from the heat source to the material (powder) and the conduction of the heat to the already solidified layers. In most of the studies for simplification reasons, radiation losses

Fig. 6 Modelling of the process

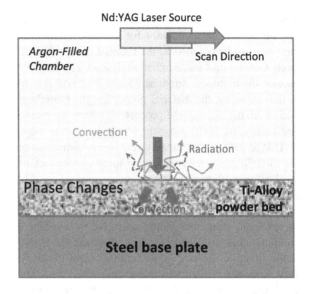

from the free surfaces [33] and heat loss from the bottom of the substrate [34] are neglected. Such simplifications regarding heat transfer phenomena can reduce the accuracy of the analysis. However, loss of accuracy occurs only to a limited extent as the main heat transfer mechanism is conduction through the powder bed and solidified regions. In only few studies are radiation losses considered for improving the accuracy of the predictions [35–39].

Considering the three dimensions for modelling the AM process, the heat conduction equation according to Fourier's law is described as

$$\frac{\partial}{\partial x}\left(k\frac{\partial T}{\partial x}\right) + \frac{\partial}{\partial y}\left(k\frac{\partial T}{\partial y}\right) + \frac{\partial}{\partial z}\left(k\frac{\partial T}{\partial z}\right) + \dot{q} = \rho C_p \frac{\partial T}{\partial t} \qquad (2)$$

where k is the thermal conductivity, T is the temperature of the part, $\dot{q}$ is the heat input rate supplied to the component, ρ is the density of the material, C_p is the specific heat capacity, and t is the interaction time between the beam and the material.

In order to consider the solid–liquid transformation, the above equation needs to be modified for taking into account the enthalpy change due to the phase transformation. The enthalpy change dH is given by:

$$dH = C_p dT \qquad (3)$$

The heat conduction Eq. (2) is thus transformed into:

$$\frac{\partial}{\partial x}\left(k\frac{\partial T}{\partial x}\right) + \frac{\partial}{\partial y}\left(k\frac{\partial T}{\partial y}\right) + \frac{\partial}{\partial z}\left(k\frac{\partial T}{\partial z}\right) + \dot{q} = \rho \frac{\partial H}{\partial t} \qquad (4)$$

The effect of phase changes can also be included in the model by considering the latent heat required for phase change at the melting point [40–42].

The initial conditions of the heat transfer problem can be expressed as

$$T(x, y, z, 0) = T_0 \tag{5}$$

This equation is based on the assumption that the initial temperature of the powder is equal to the ambient temperature T_0. However, in many cases in real AM machines the metallic powder is preheated; this has a great impact on the energy efficiency of the process as discussed in the following sections.

In the general case where the heat supplied by the beam is modelled as a heat flux and heat is lost from the system by convection and radiation, the boundary condition for the free surfaces of the part is:

$$k\frac{\partial T}{\partial n} - \dot{q}_s + h(T - T_0) + \sigma\varepsilon(T^4 - T_0^4) = 0 \tag{6}$$

where n is the vector normal to the surface, $\dot{q}_s$ is the rate of the heat input from the beam, h is the heat transfer coefficient, σ is the Stefan–Boltzmann constant, and ε is the emissivity.

The boundary conditions reflect the fixing of the part on the bed table and the heat input source. The impact of the heat source in the powder bed can be modelled as a heat flux load or as a temperature load. A number of different intensity distributions have been used in the literature up to now. In most of the studies, the heat flux source follows a Gaussian intensity distribution. The intensity $I(r)$ of a Gaussian heat flux at a radial distance from the beam centre equal to r is given by the following equation [40],

$$I(r) = \frac{2AP}{\pi\omega^2}\exp\left(-\frac{2r^2}{\omega^2}\right) \tag{7}$$

where A is the absorptivity of the powder, P is the beam power, and ω is the radius where the intensity is reduced from the intensity at the centre of the beam by a factor of e^2.

Alternatively, the heat flux has been modelled in some studies as having a constant power distribution [43, 44]. In both these cases the scanning of the laser beam is considered. Alternatively, for simplifying the analysis and saving computational effort and time, the movement of the laser source can be neglected and assumed that each layer is built as one, within a single time step [45]. A further simplification to this approach is to model the heat source as a temperature load [46, 47] with a temperature equal to that of the powder melting point. The three different approaches have been compared by Salonitis et al. [45] and presented graphically in Fig. 7.

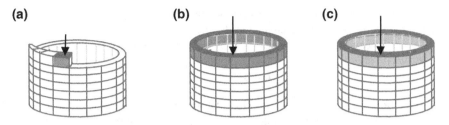

Fig. 7 Three different approaches to loading the model [45]. **a** Loaded element with Q. **b** Loaded element with Q_{ent}. **c** Consider T_{melt} fort $= t_{layer}$

3.2 Modelling of the Structural Problem During AM

As mentioned, one of the key challenges with AM is to predict the final deformation of the manufactured part and the distribution of the residual stresses. The residual stresses are the result of inhomogeneous plastic deformations during heating and cooling of the powder. The generation of the residual stresses is quite complex, with numerous affecting factors. The material type is one of the most important ones with heat transfer coefficient, thermophysical and mechanical properties, and phase composition greatly influencing the residual stresses. The higher the yield strength of the material, the more elastic the thermal and transformation induced macroscopic stresses will be generated in the part to be quenched. Thus, the residual stresses in general will be lowered with increasing yield strength of the material. Schoinochoritis et al. [11] reviewed the studies that have been presented with regard to the simulation of metallic AM processes. They have concluded that most of the studies predicting the residual stresses and distortion agree that high tensile stresses are observed at the top layer and high compressive stresses at the interface between the part and the substrate. Thin-walled structures were found to exhibit high residual stresses. Two key mechanisms lead to the generation of residual stress in the AM process [48]. The temperature gradient mechanism that is caused by the significant thermal gradient around the laser spot is considered to be the first one. The high heat input at the engagement area (thus on the top layer of the manufactured part) combined with the relatively low heat conduction beneath results in a large thermal gradient, thus the expansion of the top layers is more severe than that of the bottom ones and elastic compressive strains are generated. Furthermore, plastic strains can be induced if the yield strength of the material is reached. After cooling and shrinking on the top layers, the elastic compressive strains tend to disappear and the plastic ones remain, thus residual stresses are generated. The second mechanism works when the top layers are melted. As the melted material is cooling down, shrinking on the top layers appears. However, this deformation is restrained by the layers beneath, thus tensile stresses are induced at the top layer and compressive stresses below. Salonitis et al. [45] used finite element analysis and the level-set method for predicting the residual stresses during hybrid AM processing (in that case laser cladding followed by high-speed machining).

3.3 Numerical Analysis

The system of equations presented thus far is overly complicated to be solved analytically, and therefore numerical methods need to be used. These can be classified into mesh-based methods and mesh-free methods. The finite element analysis (FEA) method belongs to the first class. For using FEA a continuum of matter (domain) is discretised into a finite number of elements forming a mesh, thus reducing the problem to that of a finite number of unknowns. Each finite element possesses simpler geometry and therefore it is easier to analyse than the actual structure. FEA has proved to be very popular in the study of the physical phenomena during additive manufacturing processes and their parametric optimisation.

Going back to the problem described thus far, the heat transfer problem (Eq. (4)) is described by the following equation,

$$[C(T)]\{T'(t)\} + [K(T)]\{T(t)\} + \{v\} = \{Q(t)\} \tag{8}$$

where $[K]$ is the conductivity matrix, $[C]$ the specific heat matrix; $\{T\}$ the vector of nodal temperatures; $\{T'\}$ the time derivative vector of $\{T\}$; $\{v\}$ is the velocity vector, which is equal to zero as no mass transport is assumed in the current problem; and $\{Q\}$ is the nodal heat flow vector.

Similarly, the structural problem needs to be solved. The coupling of the two problems is achieved by entering the thermal results to the structural model. Thermal elements are replaced with elastic–plastic elements. The resulting model undergoes a nonlinear elastic–plastic structural analysis using temperature-dependent material properties and a multilinear isotropic hardening model. The nonlinear mechanical analysis problem is described by the following general finite element equation,

$$[K(T)]\{u(t)\} + \{F(t)\} + \{F^{th}(t)\} = 0 \tag{9}$$

where $[K(T)]$ is the temperature-dependent stiffness matrix, $\{F(t)\}$ is the external load vector, $\{F^{th}(t)\}$ is the temperature load vector, and $\{u(T)\}$ is the displacement vector.

For each load step, the nodal temperatures from the thermal analysis are read into the structural analysis. Nodal temperatures from thermal results are continued to be read into the structural analysis until the time when the model temperature has reached the ambient one.

For the sake of completion, it should be noted that FEM is not the only numerical method that can be used for such a problem. In a number of studies methods such as the Lattice–Boltzmann method (LBM) [49–52] and the finite volume method (FVM) [53, 54], have been used. In the latter ones, FVM has been used in order to take into consideration the fluid dynamics of the melt pool.

3.4 Other Considerations for the Numerical Modelling of the Process

For the numerical solution of the equations described thus far, a number of decisions are required such as whether the problem will be solved into two or three dimensions, the meshing of the geometry (element type, size, density, etc.), the material properties, and so on.

As mentioned, the process modelling and simulation can performed in only two dimensions [55–58] in order that the computational time is reduced or is three-dimensional [33, 59, 60] for higher precision and accuracy. Two-dimensional modelling can be used when melting of a single layer is simulated, but is inadequate when multiple layers are processed.

The characteristics and the quality of the meshing of the geometry are of paramount importance for the accuracy of the predictions. Furthermore, the mesh quality has a remarkable impact on the computational efficiency of the simulation and the time required for the analysis. In most of the modelling studies (regardless of the simulated manufacturing process), the mesh density is varied along the whole model considering where a higher level accuracy is expected and needed. This is obviously the case in modelling AM as well [61]. Certain regions, particularly those close to the heat source, exhibit steep temperature gradients and an accurate analysis would require a more dense mesh in those areas. As indicated by Schoinochoritis et al. [11], local refinement techniques are used for updating the mesh density in each load step.

One particular challenge with regard to the modelling of the AM process has to do with the fact that the material is gradually added to the geometry. For the modelling of such a problem, the element birth and death technique is used to simulate the process of adding material in the present study. Schoinochoritis et al. [11] indicated that the material in powder or liquid form does not contribute to the overall stiffness of the model; this fact can be used for improving the accuracy of the predictions. The transition of the material from powder to liquid and finally to solid, can be used based on this method. The elements are deactivated until solidification. Although they are visually present in the model, they do not add to the overall stiffness of the matrix [33, 62] as their actual stiffness has been multiplied by a severe reduction factor. If we consider Eq. (9), this means that the elements that are not yet activated in the matrix $[K(T)]$ are multiplied by the reduction factor. After solidification the elements are activated, regaining their actual stiffness. A control loop determines when an element is activated according to whether its temperature after a load step has surpassed the melting point [42]. As mentioned in Fig. 6, the birth of the elements can take place either one by one, or the whole layer simultaneously [45]. From a programming point of view, controlling the state of the elements and locally refining the mesh in each load step increases the level of difficulty in the preparation of the analysis.

Finally, the predictions of the modelling are affected by the quality of the material properties that are in hand. Inasmuch as during the process the temperature levels exceed that of the melting point, the material properties should be defined as a function of temperature, considering, however, as well the transition of the material from powder to liquid and finally to solid. The properties that affect significantly the process evolution and their dependence on temperature is of high interest and are the density, the thermal conductivity, and the specific heat capacity [11]. Other material properties, such as the Young modulus, the tangent modulus, the yield strength, and the thermal expansion coefficient, can also be considered as temperature dependent. Those properties are used when a mechanical analysis is performed. Thus, their temperature dependency affects the residual stresses and distortion results. The powder particles can be modelled explicitly [63] or represented by a continuum body with adjusted properties [64], which is isotropic and homogeneous. Modelling explicitly the powder particles can offer increased accuracy and microscopic insight, but has serious effects on the computational effort required. Schoinochoritis et al. [11] analysed the effect of the various material properties and the impact of considering them either as temperature—dependent or independent ones.

4 Improving Energy Efficiency of Powder Bed AM Processes Using Modelling

4.1 Introduction

The energy efficiency analysis can take place on different levels depending on the scope of the analysis. As indicated by Duflou et al. [65], five different levels can be identified, namely:

- Device/process level
- Line/cell/multimachine system
- Facility
- Multifactory system
- Enterprise/global supply chain

Each of these analysis levels relies on different assumptions, different input, and provides different results. All the levels can be affected by the AM processes and it makes sense to analyse such processes from their perspective. With regard to conventional manufacturing processes, a number of recent studies have been published dealing with energy efficiency, however, most of these studies rely either solely on the monitoring of the energy consumption of machine tools [66, 67] or on the monitoring of specific machine tool components, such as the spindle [68].

The energy efficiency is linked to the energy consumed by the manufacturing process (therefore the machine tool/additive manufacturing equipment and the auxiliary devices). As shown in Sect. 2, the energy consumption is directly related to two out of the three sustainability dimensions (economic and environmental). It can, however, be related indirectly to the social dimension as well. Chen et al. [19]

compared the energy requirements of a number of alternative AM processes (such as selective laser sintering, fused deposition modelling, and 3D printing) with injection moulding (obvious for plastic products). The comparison was based on the specific energy demand per weight unit. It was shown that the energy consumption from the literature indicated a large variation depending on a number of factors such as the process parameters. The scanning speed/process rate was found to be the driving process parameter with regard to the energy consumption. An interesting finding was that selective laser sintering, which is a powder-bed–based AM process, is significantly more energy intensive than injection moulding but at the same time significantly less intensive than other AM processes. However, as indicated by a number of researchers [19, 22, 69], the energy demand of the process itself is just one perspective; other effects need to be taken into consideration as well inasmuch as they indirectly influence the energy demand of DDM from the system perspective. Chen et al. [19] identified the key ones being the material yield (i.e., the ratio of final product weight and necessary input material weight), the need for tools and moulds, the complexity and individuality of parts, and the possible supply chain and transportation effects. For addressing these problems, the idea of using the embodied energy (the accumulated energy consumed for the production of any product, considered as if that energy were incorporated or 'embodied' in the product itself) was adapted. It was shown that at the lower batch size, selective laser sintering has lower embodied energy than that of the injection moulding, however, this is reversed as the batch size is increased.

4.2   Improving the Energy Efficiency of the Process

Before deciding a strategy for the energy efficiency optimisation of the process, a thorough energy consumption audit is required. For the case of conventional manufacturing processes (such as machining or grinding), the analysis relies on the energy audit of the machine tool during the processing. During the last years, a number of studies have been presented dealing with the energy efficiency at this level, indicatively the 'unit process energy' method [70] and the 'energy blocks' method [71] that have been already discussed. The energy consumed by machine tools during machining is significantly greater than the theoretical energy required in chip formation. As an example, the specific cutting energy accounts for less than 15 % of the total energy consumed by a modern automatic machine tool during machining [72, 73]. For the determination of the energy consumption of the peripherals of the machine tools, the monitoring procedure has to be designed thoroughly in advance. Salonitis [74] developed a framework for determining the energy consumption of a machine tool that can be adapted for the needs of the additive manufacturing machine as well.

Measuring the energy consumption of machine tools (and in the present case additive manufacturing machines) poses a number of challenges with the main one being that when measuring the consumption of a machine, a number of subsystems and peripherals are present working simultaneously that cannot be isolated and measured individually. The framework presented by Salonitis for addressing this problem is composed of three major phases: the preparation phase, the measurement phase, and the analysis phase. Within the preparation phase, the energy audit approach is structured and designed based on the characteristics of the machine tool to be analysed. Within the second phase all the measurements are taking place. The final phase deals with the analysis of the results. The framework is adapted for the selective laser sintering process and schematically presented in Fig. 8. After measuring the energy consumption during the process, the energy consumed from each component can be estimated. Using Pareto analysis the various subsystems are ranked with regard to energy consumption, establishing in this way which subsystems are best to focus improvement efforts.

However, because the framework has not yet been applied for the energy audit of an AM process, it was necessary to look for data from the literature. In powder bed AM processes (such as the SLS process) a number of components of the machine used for the process consume a considerable amount of energy. If we consider, for example, laser sintering, the material is in powder format, typically with an average grain diameter from 20–50 μm. After a powder layer (normally 0.1 mm) is laid on the building platform, a laser beam scans the powder bed tracing the layer geometry. Subsequently, the building platform is moved down equal to the chosen layer thickness and the feed container moves upward by a defined height. A roller puts a new layer homogeneously on the already solidified layer. This process continues until the parts are completely produced. Finally the parts cool down, before it can be cleaned. To improve several properties, such as the surface of the part, different postprocessing steps are necessary. In powder bed AM processes, preheating of the part and the powder prior to the solidification of a layer is used to reduce thermal residual stresses in the final part which are a severe quality issue in AM processes [75].

Sreenivasan et al. [76] experimentally estimated the energy consumption from all components of an SLS Vanguard[TM] HiQ + MS machine that was manufactured by 3D Systems Corp. They measured the total power input to the machine across all three phases of the connection. The results of their analysis are shown in Fig. 9. The heater system that is used for heating the powder bed is considered to be the largest accumulator of electricity draw (accounting for almost 40 % of the total power), followed by the stepper motor system which controls the piston motion of the powder bed, then the roller system which spreads the powder across the bed, and finally the laser system. These results are in agreement with Reinhardt and Witt's [77] findings as well on a different SLS machine (DTM 2000 laser sintering machine).

According to Fig. 8 and the two aforementioned studies almost 40 % of the energy in AM powder-based processes is consumed during preheating. Thus, increasing the energy efficiency of preheating can have a significant impact on the overall increase of the energy efficiency of powder bed AM processes.

The energy consumption of the AM process and how to optimise it has been addressed in a number of studies. Baumers et al. measured the energy consumption

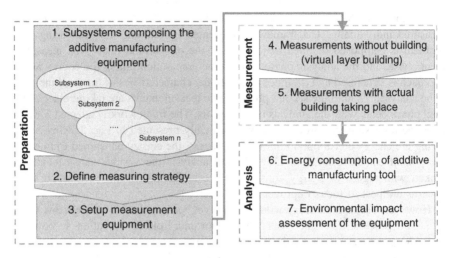

Fig. 8 Energy consumption measurement framework (based on [74, 87])

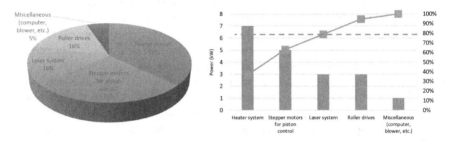

Fig. 9 Energy consumption of individual components; *left* subsystem energy demand; *right* Pareto analysis (based on experimental results reported by Sreenivasan et al. [76])

of 3D Systems' Sinterstation HIQ + HS and the EOS GmbH EOSINT P390 laser sintering machines [78]. The total energy consumption was shown that depends on the build time and therefore increasing the process speed would positively influence both its energy efficiency and the process economics as well. Similar analyses have been performed for other AM processes as well, such as the bind-jetting process [79] and direct metal deposition [80]. The build time is a function of the volume of the part, the process parameters selected, and the orientation of the part. The volume of the part is a design characteristic and in most cases it is a 'given' one that cannot be altered. However, both the process parameters and the building orientation can be optimised having in mind the energy efficiency of the process. Thus in a number of studies, this problem has been approached through different routes such as using the interior-point algorithm for selecting the optimal part orientation and layer thickness for minimising the energy consumption in the SLS process [81], or using the genetic algorithm [82] or Pareto-based approach [83].

Few studies have been presented employing modelling tools for assessing the energy consumption of machine tools, manufacturing processes, and specifically AM. One example is to model environment as a thermodynamic system, an approach employed by Gutowski et al. [84]. The main challenge for using this approach lies in that it results in a very complex energy problem. This difficulty can be simplified with the use of exergy or 'available work'. However, the reliability of such an approach relies in extensive experimentation.

4.3 Preheating Modelling

Going back to the initial observation that the preheating stage is the key energy consumer and that focusing on optimising this stage would potentially result in the highest benefits, it was found that only few studies focused on this aspect. The literature review indicated that the preheating energy in SLS is proportional to the number of the sintered layers [78, 83]. Because the build orientation of the part determines the z-height of the part and thus the number of layers, for a given layer thickness, preheating energy is directly influenced by the build orientation.

There are three preheating methods that are widely used in the powder-based AM processes depending on whether the heating is focused on the area where the powder is to be solidified or whether the whole part is heated. The former case is when a laser is used preheating the whole chamber [85] or just the heating of the base plate [75]. These methods are presented in Fig. 10.

Papadakis et al. [86] compared the three methods using finite element analysis for the case of a RENISHAW AM250 machine manufacturing a TiAl6V4 cantilever beam. Figure 11 illustrates the results of the energy consumption in joules of individual virtual layers for the three preheating methods. This consideration initially regards the energy spent into the actual twin cantilever volume and does not consider the surrounding preheating of the surrounding powder which has not been

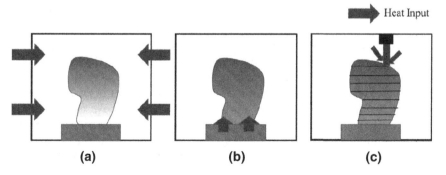

Fig. 10 The three preheating methods under investigation: (**a**) chamber, (**b**) base plate, and (**c**) laser prescanning heating [86]

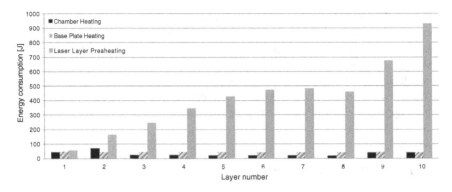

Fig. 11 Energy losses per layer for the three preheating methods [86]

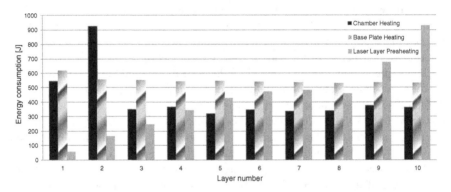

Fig. 12 Energy losses per layer for the three preheating methods in a specific powder bed of area [86]

melted during SLM processing; that is, powder is theoretically only deposited in the area of the actual twin cantilever geometry. However, considering the heat flux calculations for the actual model volumes a deduction for a proposed powder bed of the specific area is performed for the preheating of the whole chamber and of the base plate. Figure 12 shows the results of the individual model layers for a specific powder bed surface (despite the fact that the base plate surface is significantly larger), so that comparable energy consumptions for the three preheating methods result. It is noted that for the case of prescanning heating there were no changes in the consumed energy due to the selective preheating of the laser in the areas where the material was solidified and the structure was built. The above computational findings concerning energy usage during preheating of the powder bed and the part under fabrication, in SLM/SLS, provide a very helpful basis for selecting the most appropriate preheating method. Initially, it is observed that for the preheating methods (i) and (ii) the energy consumption per layer is relatively independent of the geometry of the part. However, the laser prescanning heating method (iii) is very strongly dependent, in terms of energy consumption, on the volume to be scanned in each layer, that is, the component geometry.

5 Conclusion and Outcome

The present chapter presented and discussed an overview of powder bed additive manufacturing implications to sustainability and energy efficiency. Relevant recent studies have been reviewed. It is obvious that AM has a very promising future, although it is not expected that AM will replace conventional manufacturing processes, but rather complement them for specific batch sizes and customised products.

The implications to the sustainability were discussed and sustainability key performance indicators that have been presented in the past were collected, listed, and discussed. This chapter also presented a comprehensive review of the relevant literature of the studies presented on the energy efficiency of metallic powder bed additive manufacturing processes and the key challenges for improving it.

Additionally, the basics for modelling the process with finite element simulation were discussed for the estimation of the energy efficiency and the optimisation of the process under this prism. Because the preheating of the raw material is the key energy consumer, the alternative methods were discussed and compared.

References

1. Kruth J-P, Leu MC, Nakagawa T (1998) Progress in Additive manufacturing and rapid prototyping. CIRP Ann Manufact Technol 47(2):525–540
2. Wohlers (2013) Wohlers Report: additive manufacturing and 3D printing state of the industry. Wohlers Associates, USA
3. RAEng (2013) Additive manufacturing: opportunities and constraints. Royal Academy of Engineering, UK
4. Levy GN, Schindel R, Kruth JP (2003) Rapid manufacturing and rapid tooling with layer manufacturing (LM) technologies, state of the art and future perspectives. CIRP Ann Manuf Technol 52(2):589–609
5. Levy GN, Schindel R, Kruth JP (2003) Rapid manufacturing and rapid tooling with layer manufacturing (LM) technologies, state of the art and future perspectives. CIRP Ann Manuf Technol 52(2):589–609
6. Santos EC, Shiomi M, Osakada K, Laoui T (2006) Rapid manufacturing of metal components by laser forming. Int J Mach Tool Manuf 46(12–13):1459–1468
7. Gu DD, Meiners W, Wissenbach K, Poprawe R (2012) Laser additive manufacturing of metallic components: materials, processes and mechanisms. Int Mater Rev 57(3):133–164
8. Leuders S, Thöne M, Riemer A, Niendorf T, Tröster T, Richard HA, Maier HJ (2013) On the mechanical behaviour of titanium alloy TiAl6V4 manufactured by selective laser melting: fatigue resistance and crack growth performance. Int J Fatigue 448:300–307
9. Kruth JP, Levy G, Klocke F, Childs THC (2007) Consolidation phenomena in laser and powder-bed based layered manufacturing. CIRP Ann Manuf Technol 56(2):730–759

10. Gu D, Wang H, Zhang G (2014) Selective laser melting additive manufacturing of Ti-based nanocomposites: the role of nanopowder. Metall Mater Trans A 45(1):464–476
11. Schoinochoritis B, Chantzis D, Salonitis K (2015) Simulation of metallic powder bed additive manufacturing processes with the finite element method: a critical review. Proc Instit Mech Eng Part B J Eng Manuf. doi: 10.1177/0954405414567522
12. IEA (2007) Tracking industrial energy efficiency and CO2 emission
13. Brundtland GH (1987) Our common future. Oxford University Press, Oxford
14. Mani M, Lyons KW, Gupta SK (2014) Sustainability characterization for additive manufacturing. J Res National Inst Stand Technol 119:419–428
15. Ellen McArthour Foundation (2015) Accessible at http://www.ellenmacarthurfoundation.org/. Accessed on 20 Nov 2015
16. Salonitis K, Stavropoulos P (2013) On the integration of the CAx systems towards sustainable production. Procedia CIRP, vol 9, pp 115–120. doi:10.1016/j.procir.2013.06.178
17. Gebler M, SchootUiterkamp AJM, Visser C (2014) A global sustainability perspective on 3D printing technologies. Energy Policy 74(C):158–167
18. Despeisse M, Ford S (2015) The role of additive manufacturing in improving resource efficiency and sustainability. In: Proceedings of the APMS 2015 international conference
19. Chen D, Heyer S, Ibbotson S, Salonitis K, GarðarSteingrímsson J, Thiede S (2015) Direct digital manufacturing: definition, evolution, and sustainability implications. J Cleaner Prod 107:615–625
20. Baumers M, Tuck C, Wildman R, Ashcroft I, Rosamond E, Hague R (2012) Transparency built-in energy consumption and cost estimation for additive manufacturing. J Ind Ecol 17 (3):418–431
21. Atzeni E, Salmi A (2012) Economics of additive manufacturing for end-usable metal parts. Int J Adv Manuf Technol 62(9):1147–1155
22. Sreenivasan R, Goel A, Bourell DL (2010) Sustainability issues in laser-based additive manufacturing. Phys Procedia 5:81–90
23. Serres N, Tidu D, Sankare S, Hlawka F (2011) Environmental comparison of MESO-CLAD process and conventional machining implementing life cycle assessment. J Cleaner Prod 19:1117–1124
24. Seuring S (2004) Integrated chain management and supply chain management comparative analysis and illustrative cases. J Cleaner Prod 12:1059–1071
25. Kleindorfer PR, Singhal K, Van Wassenhove LN (2005) Sustainable operations management. Prod Oper Manage 14(4):482–492
26. Linton JD, Klassen R, Jayaraman V (2007) Sustainable supply chains: an introduction. J Oper Manage 25:1075–1082
27. Bell S, Morse S (2008) Sustainability Indicators—measuring the immeasurable? ISBN-13: 978-1-84497-299-6, published by Earthscan, UK, p 223
28. Stiglitz and Sen—Fitoussi report (2009)—Report by the commission on the measurement of economic performance and social progress (French Government Initiative)
29. Huang SH, Liu P, Mokasdar A, Hou L (2013) Additive manufacturing and its societal impact: a literature review. Int J Adv Manuf Technol 67:1191–1203
30. Chryssolouris G (2006) Manuf Syst Theory Pract, 3rd edn. Springer-Verlag, New York
31. Salonitis K, Ball P (2013) Energy efficient manufacturing from machine tools to manufacturing systems. Procedia CIRP 7:634–639
32. Bunse K, Vodicka M, Schönsleben P, Brülhart M, Ernst FO (2011) Integrating energy efficiency performance in production management—gap analysis between industrial needs and scientific literature. J Cleaner Prod 19:667–679
33. Roberts IA, Wang CJ, Esterlein R et al (2009) A three-dimensional finite element analysis of the temperature field during laser melting of metal powders in additive layer manufacturing. Int J Mach Tool Manuf 49:916–923
34. Zhang J, Li D, Li J et al (2011) Numerical simulation of temperature field in selective laser sintering. In: Li D, Liu Y, Chen Y (eds) Computer and computing technologies in agriculture IV, 1st edn. Springer, New York, pp 474–479

35. Bai PK, Cheng J, Liu B et al (2006) Numerical simulation of temperature field during selective laser sintering of polymer-coated molybdenum powder. T Nonferr Metal Soc 16:603–607
36. Van Belle L, Vansteenkiste G, Boyer JC (2012) Comparisons of numerical modelling of the selective laser melting. Key Eng Mat 504–506:1067–1072
37. Zhang DQ, Cai QZ, Liu JH et al (2010) Select laser melting of W-Ni–Fe powders: simulation and experimental study. Int J Adv Manuf Technol 51:649–658
38. Song B, Dong S, Liao H et al (2012) Process parameter selection for selective laser melting of Ti6Al4 V based on temperature distribution simulation and experimental sintering. Int J Adv Manuf Technol 61(9–12):967–974
39. Yin J, Zhu H, Ke L et al (2012) Simulation of temperature distribution in single metallic powder layer for laser micro-sintering. Comput Mater Sci 53(1):333–339
40. Hussein A, Hao L, Yan C et al (2013) Finite element simulation of the temperature and stress fields in single layers built without-support in selective laser melting. Mater Des 52:638–647
41. Patil RB, Yadava V (2007) Finite element analysis of temperature distribution in single metallic powder layer during metal laser sintering. Int J Mach Tool Manuf 47:1069–1080
42. Shen N, Chou K (2012) Thermal modeling of electron beam additive manufacturing process— powder sintering effects. In: ASME international manufacturing science and engineering conference, Notre Dame, USA, 4–8 June 2012, pp 287–295
43. Contuzzi N, Campanelli SL, Ludovico AD (2011) 3D Finite element analysis in the selective laser melting process. Int J Simul Model 10(3):113–121
44. Matsumoto M, Shiomi M, Osakada K et al (2002) Finite element analysis of single layer forming on metallic powder bed in rapid prototyping by selective laser processing. Int J Mach Tool Manuf 42(1):61–67
45. Salonitis K, D'Alvice L, Schoinochoritis B, Chantzis D (2015) Additive manufacturing and post-processing simulation: laser cladding followed by high speed machining. Int J Adv Manuf Technol. doi:10.1007/s00170-015-7989-y
46. Schilp J, Seidel C, Krauss H, et al. (2014) Investigations on temperature fields during laser beam melting by means of process monitoring and multiscale process modelling. Adv Mech Eng 6
47. Ma L, Bin H (2007) Temperature and stress analysis and simulation in fractal scanning-based laser sintering. Int J Adv Manuf Technol 34(9–10):898–903
48. Mercelis P, Kruth J (2006) Residual stresses in selective laser sintering and selective laser melting. Rapid Prototyping J 12(5):254–265
49. Bauereiß A, Scharowsky T, Körner C (2014) Defect generation and propagation mechanism during additive manufacturing by selective beam melting. J Mater Process Technol 214 (11):2522–2528
50. Körner C, Bauereiß A, Attar E (2013) Fundamental consolidation mechanisms during selective beam melting of powders. Model Simul Mater Sci Eng 21(8):085011
51. Körner C, Attar E, Heinl P (2011) Mesoscopic simulation of selective beam melting processes. J Mater Process Technol 211(6):978–987
52. Scharowsky T, Bauereiß A, Singer RF, et al. Observation and numerical simulation of melt pool dynamic and beam powder interaction during selective electron beam melting. In: National science foundation solid freeform fabrication symposium, Austin, USA, 6–8 August 2012
53. Chen T, Zhang Y (2004) Numerical simulation of two dimensional melting and resolidification of a two-component metal powder layer in selective laser sintering process. Numer Heat Tr A-Appl 46(7):633–649
54. Dai D, Gu D (2014) Thermal behavior and densification mechanism during selective laser melting of copper matrix composites: simulation and experiments. Mater Des 55:482–491
55. Matsumoto M, Shiomi M, Osakada K et al (2002) Finite element analysis of single layer forming on metallic powder bed in rapid prototyping by selective laser processing. Int J Mach Tool Manuf u 42(1):61–67
56. Van Belle L, Vansteenkiste G, Boyer JC (2012) Comparisons of numerical modelling of the selective laser melting. Key Eng Mat 504–506:1067–1072

57. Bai PK, Cheng J, Liu B et al (2006) Numerical simulation of temperature field during selective laser sintering of polymer-coated molybdenum powder. T Nonferr Metal Soc 16:603–607

58. Childs THC, Hauser C, Taylor CM, et al. (2000) Simulation and experimental verification of crystalline polymer and direct metal Selective Laser Sintering. In: National science foundation solid freeform fabrication symposium, Austin, USA, 7–9 August 2000

59. Kolossov S, Boillat E, Glardon R et al (2004) 3D FE simulation for temperature evolution in the selective laser sintering process. Int J Mach Tool Manuf 44:117–123

60. Contuzzi N, Campanelli SL, Ludovico AD (2011) 3D Finite element analysis in the selective laser melting process. IJSIMM 10(3):113–121

61. Riedlbauer D, Steinmann P, Mergheim J (2014) Thermomechanical finite element simulations of selective electron beam melting processes: performance considerations. Comput Mech 54 (1):109–122

62. Denlinger ER, Heigel JC, Michaleris P (2014) Residual stress and distortion modeling of electron beam direct manufacturing Ti-6Al-4 V. Proc Inst Mech Eng Part B J Eng Manuf. doi: 10.1177/0954405414539494

63. Liu FR, Zhang Q, Zhou WP et al (2012) Micro scale 3D FEM simulation on thermal evolution within the porous structure in selective laser sintering. J Mater Process Tech 212(10):2058–2065

64. Schilp J, Seidel C, Krauss H, et al. (2014) Investigations on temperature fields during laser beam melting by means of process monitoring and multiscale process modelling. Adv Mech Eng 6

65. Duflou JR, Sutherland JW, Dornfield S, Herrmann C, Jeswiet J, Kara S, Hauschild M, Kellens K (2012) Towards energy and resource efficient manufacturing: a processes and systems approach. CIRP Ann Manuf Technol 61:587–609

66. Vijayaraghavan A, Dornfeld D (2010) Automated energy monitoring of machine tools. CIRP Ann Manuf Technol 59:21–24

67. Behrendt T, Zeina A, Min S (2012) Development of an energy consumption monitoring procedure for machine tools. CIRP Ann Manuf Technol 61:43–46

68. Oda Y, Mori M, Ogawa K, Nishida S, Fujishima M, Kawamura T (2012) Study of optimal cutting condition for energy efficiency improvement in ball end milling with tool-workpiece inclination. CIRP Ann Manuf Technol 61:119–122

69. Hao L, Raymond D, Strano G, Dadbakhsh S (2010) Enhancing the sustainability of additive manufacturing. In: ICRM2010-Green Manufacturing, Ningbo, China, pp 390–395

70. Kara S, Li W (2011) Unit process energy consumption models for material removal processes. CIRP Ann Manuf Technol 60:37–40

71. Weinert N, Chiotellis S, Seliger G (2011) Methodology for planning and operating energy-efficient production systems. CIRP Ann Manuf Technol 60:41–44

72. Dahmus J, Gutowski T (2004) An environmental analysis of machining. In: Proceedings of ASME international mechanical engineering congress and R&D exposition, pp 13–19

73. Salonitis K (2012) Efficient grinding processes: an energy efficiency point of view. In: Proceedings of the 10th international conference on manufacturing research (ICMR 2012), pp 541–546

74. Salonitis K (2015) Energy efficiency assessment of grinding strategy. Int J Energy Sect Manage 9(1):20–37

75. Kempen K, Thijs L, Vrancken B, Van Humbeeck J, Kruth JP (2013) Producing crack-free, high density M2 Hss parts by selective laser melting: pre-heating the baseplate. In: Proceedings of the 24th international solid freeform fabrication symposium. Laboratory for freeform fabrication, Austin, TX, pp 131–139

76. Sreenivasan R, Goel A, Bourell DL (2010) Sustainability issues in laser-based additive manufacturing. Phys Procedia 5:81–90. doi:10.1016/j.phpro.2010.08.124

77. Reinhardt T, Witt G (2012) Experimental analysis of the laser-sintering process from an energetic point of view. Ann. DAAAM 2012 Proc. 23rd Int. DAAAM Symp 23:405–408

78. Baumers M, Tuck C, Bourell DL, Sreenivasan R, Hague R (2011) Sustainability of additive manufacturing: measuring the energy consumption of the laser sintering process. Proc Inst Mech Eng Part B J Eng Manuf 225:2228–2239
79. Meteyer S, Xu X, Perry N, Zhao YF (2014) Energy and material flow analysis of binder-jetting additive manufacturing processes. Procedia CIRP 15:19–25. doi:10.1016/j. procir.2014.06.030
80. Morrow WR, Qi H, Kim I, Mazumder J, Skerlos SJ (2007) Environmental aspects of laser-based and conventional tool and die manufacturing. J Cleaner Prod 15:932–943
81. Paul R, Anand S (2012) Process energy analysis and optimization in selective laser sintering. J Manuf Syst 31:429–437
82. Verma A, Rai R (2013) Energy efficient modeling and optimization of additive manufacturing processes. In: Proceedings of the 24th international solid freeform fabrication symposium. Laboratory for Freeform Fabrication, Austin, TX, pp 231–241
83. Strano G, Hao L, Everson RM, Evans KE (2011) Multi-objective optimization of selective laser sintering processes for surface quality and energy saving. Proc Inst Mech Eng Part B J Eng Manuf 225:1673–1682
84. Gutowski T, Dahmus J, Thiriez A (2006) Electrical energy requirements for manufacturing processes. In 13th CIRP international conference on life cycle engineering
85. Renaldi R, Dewulf W, Kruth J, Duflou JR (2014) Environmental impact modeling of selective laser sintering processes. Rapid Prototyping J 20:459–470. doi:10.1108/RPJ-02-2013-0018
86. Papadakis L, Schoinochoritis B, Chantzis D, Doukas C, Salonitis K (2015) On the energy efficiency of pre-heating methods in SLM/SLS processes. Working paper to be submitted for publication
87. Salonitis K (2015) Grind-hardening process. SpringerBrief, New York. doi: 10.1007/978-3-319-19372-4

Environmental Impact Assessment Studies in Additive Manufacturing

Olivier Kerbrat, Florent Le Bourhis, Pascal Mognol
and Jean-Yves Hascoët

Abstract This chapter focuses on the environmental studies in additive manufacturing. For a cleaner production, environmental impacts that occur during the manufacturing phase should be assessed with accuracy. First, the literature on all the studies led to the characterisation of the environmental impact of additive manufacturing processes. The studies on electric energy consumption of these processes are analysed here, and then some studies taking into account raw material and all the flows through the process are detailed. Secondly, a new methodology in order to evaluate, with accuracy, the environmental impact of a part from its CAD model is presented. In this methodology, the work is not focused only on electrical consumption but also on fluids and material consumption which also contribute to the environmental impact. In addition, the inputs of this methodology correspond to the set part process, which allows taking into account different manufacturing strategies and their influences on the global environmental impact. The methodology developed is based on both analytic models (validated by experiments) and experimental models. And finally, an industrial example shows that for some manufacturing strategies, the environmental impact due to electrical consumption is not the predominant one. In this case study, material consumption has an important impact and has to be taken into consideration for a complete environmental impact assessment.

Keywords Additive manufacturing · Environmental impacts · Product design optimization · Life-time performance · Electric energy consumption

O. Kerbrat (✉) · F. Le Bourhis · P. Mognol · J.-Y. Hascoët
IRCCyN, Institut de Recherche En Communications et Cybernétique de Nantes,
1 Rue de La Noë, BP 92101, 44321 Nantes, France
e-mail: Olivier.Kerbrat@irccyn.ec-nantes.fr

© Springer Science+Business Media Singapore 2016
S.S. Muthu and M.M. Savalani (eds.), *Handbook of Sustainability
in Additive Manufacturing*, Environmental Footprints and Eco-design
of Products and Processes, DOI 10.1007/978-981-10-0606-7_2

1 Introduction

This chapter gives an overall view of the environmental impact assessment applied to additive manufacturing processes. As young as the additive processes are compared to more traditional ones, the literature on this topic is relatively recent, but the number of publications drastically increases. A whole methodology to assess environmental impact is presented, with a case study in laser cladding, a directed energy deposition process. It is divided into three main sections.

The first section of this chapter is a literature review. The aim of this section is to give a precise view of what has been done when assessing environmental impact in additive manufacturing. It is divided into three subsections. In a first approach, focus is put on studies dealing with electrical energy consumption. The link between manufacturing strategies, part's orientation, process parameters, and the whole electrical energy consumption is established, based on a literature review (Luo, Mognol, Bourell, Baumers, Verna, etc.). The objectives of these studies could be to help to compare additive processes between themselves and with more 'traditional' processes (machining). At the end of this subsection, a comparative table is given to classify processes and machines considering their energy consumption rate (in KWh/kg).

Secondly, focus is put on material consumption. In fact, additive manufacturing is known to produce parts without lost material. However, a certain amount of material should be removed from the machine or the part at the end of the process. In order to reduce the environmental impact due to this lost material, a few studies (Dotchev, Gornet, etc.) try to develop methodologies to reuse (with or without new raw material) or recycle this raw material.

Finally, some studies evaluate the environmental impact considering energy, material, and fluid consumption. A few methodologies, such as the CO2PE! Initiative (Kellens, Duflou, etc.), are based on a global input–output inventory and take into account energy consumption, resource consumption, and process emissions.

Based on this state of the art, the second subsection is constituted of a whole methodology for environmental impact assessment when considering an additive process. The methodology considers the part's design and machine technology. It allows us to determine the environmental impact of the set part process. The methodology is divided into three steps: raw material preparation impact, process impact, and lost material recycling impact.

The methodology is based on predictive models that are developed to evaluate the environmental impact of the whole flux consumed (electricity, material, and fluids) during all manufacturing steps. The models concern all the features of the machine that contribute to the global environmental impact. It is a local (features)–global (impact) approach, based on an accurate modelling of the process.

Then, the third section is a case study on laser cladding, a directed energy deposition process. Fluid, material, and energy consumptions are calculated, directed from the CAD model of the part, in order to establish a predictive environmental impact assessment, during all manufacturing steps (from material extraction to powder recycling). The results can help the designers to choose the best geometry for the part when taking into consideration the environmental impact of the product in its manufacturing step.

2 Literature Review

2.1 Introduction

The first environmental studies on AM processes put forward the possibilities of gain in terms of environmental impact compared with the more traditional processes such as machining [1]. Indeed, only 10 years after the development of the first industrial AM machines, studies on the environmental impact of these processes were conducted. This was due to the necessity of taking into account these aspects, with the aim of favoring the large-scale development of the AM processes. AM already offers a new freedom of design, but their industrial development will be more important if these processes have a lesser environmental impact [2].

Five years ago, Hao et al. from the University of Exeter proposed a study allowing putting forward the possibilities offered by the AM processes to minimise electric energy consumption during manufacturing [3]. They expressed five major areas for AM to generate positive environmental impacts:

- *Material utilisation*: AM can efficiently utilise raw materials and their functionality. Nonconsolidated raw materials in a powder-based process such as powder bed fusion can be reused so that the material waste can be minimised.
- *Product design optimisation*: The free-form fabrication nature of AM enables optimisation in the design of the products. The optimal design will result in the reduction of the materials, energy, fuel, or natural resources in the product manufacturing.
- *Manufacturing process*: The AM has the potential to replace processes where significant amounts of energy are wasted, such as casting or moulding. It can also save many resources spent on the fabrication of specific tooling for production.
- *Supply chain*: As a direct digital manufacturing approach, the AM machines can be distributed closer to customers and managed by a Web-based system to coordinate the demands and requirements of product stakeholders and maximise the efficiency of the supply chains. This can reduce the need of long-distance transportation, warehousing, logistics, and, for many cases, disposable packaging.
- *Life-cycle performance*: AM can be used to repair and add advanced functions to existing products and as such the life-time performance can be extended.

In this section, we focus on the studies which led to the characterisation of the environmental impact of AM processes. In the first part, the aspects of electric energy consumption of these processes is analysed, because most of the studies deal with electric energy consumption. In the second part, we are interested in the works that are specifically on the consumption of raw material, because it is one of the main advantages of these processes. In the third part, we study the few works that take into account all the flows consumed to determine an associated environmental impact. And at the end of this section, we show the possibilities offered by AM processes on the whole life cycle of a product.

2.2 Electric Energy Consumption of AM Processes

2.2.1 First Study: Luo et al.

In a first approach, in order to estimate the environmental performance of AM processes, a number of studies were interested in their electric energy consumption. This first approach allows us to compare, on a simple criterion, the AM processes between themselves and even to compare them with the more traditional ones.

The first works led on the energy aspects were conducted by Luo et al. [4, 5]. In their studies, the authors compare three SLA machines. An equation gives the scanning speed, a second one gives the process productivity, and then the energy consumption rate (ECR, kWh/cm^3) is calculated, and the environmental impact of the energy used to process one cm^3 of epoxy resin is obtained (with eco-indicator index). The results show that the machine with the highest laser power, resulting in the highest scanning speed, has the least ECR.

These first studies are interesting because they propose a first comparison of the processes between themselves. There were completed to compare different machines by Sreenivasan et al. [6]. But these studies take into account the energy consumption of the manufacturing processes by considering only the machine, and not all the sensitive parameters (shape of the part, positioning, etc.) that can modify the ECR by modifying the power rate of the machine during the process.

2.2.2 Influence of the Manufacturing Orientation

Most of the AM processes use the concept of layer-by-layer manufacturing. This concept requires the implementation of a slicing of the part to be produced. One of the first studies taking into account the set part process to determine the electric consumption of an AM machine was proposed by Mognol et al. Indeed, to evaluate the influence of the slicing orientation of the part on the energy consumption of the machine, test parts were produced considering different manufacturing orientations and the electric energy consumption during the manufacturing has been measured.

This has been done on three technologies (material extrusion, material jetting, and powder-based fusion) [7, 8].

Figure 1 illustrates the various orientations of the part taken into account in this study. This work allows us to put forward the major influence of the manufacturing orientation on the machine consumption. This most important parameter is the total manufacturing duration, which is strongly dependent on the height to be produced. Therefore, the more important the manufacturing time is, the more important the energy consumption of the machine is.

On the same criterion of optimisation of the manufacturing orientation, Verma et al. proposed a study allowing the minimisation of the electric energy consumption and the material consumption, depending on the orientation [9]. The authors developed a multistep optimisation enabling the AM process towards energy efficiency (Fig. 2). Process objectives such as material waste and electric consumption are minimised both in the part and layer domains.

Fig. 1 The various position of the part [8]

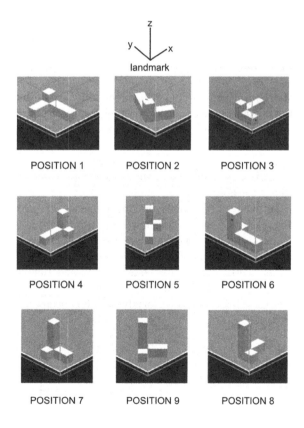

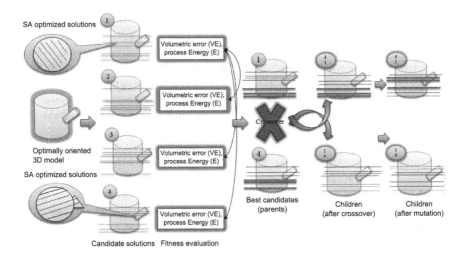

Fig. 2 Candidate solution generation and various operators on sample 3D part [9]

2.2.3 Influence of Packing Density of AM Platforms

Afterward, Baumers et al. studied the influence of the geometry of the part and the packing density of the space machine on the electric energy consumption [10, 11]. In their works, they analysed the energy consumption of two machines, one SLS and one EBM.

The part used for this study is presented in Fig. 3. The part geometry was chosen to analyse the influence of the ratio section/volume and perimeter/section on the energy consumption.

Furthermore, by analysing the influence of packing density of the platform (Fig. 4) on the energy consumption, the authors show that the consumption is not linked to the number of parts realised. It confirms the other studies showing the energy consumption is strongly dependent on the height of the part.

The works realised by Mognol et al. and Baumers et al. are very interesting because they highlight the importance of the consideration of the set part process within the framework of an analysis of the electric energy consumption of the AM processes.

Fig. 3 The standardised test part [11]

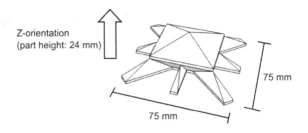

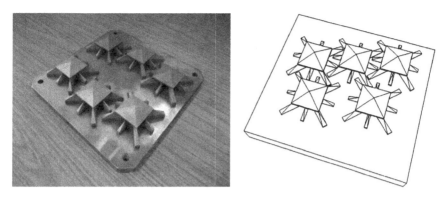

Fig. 4 Full build configuration for SLM and DMLS (*left*) and EBM (*right*) [11]

2.2.4 Comparison Between AM Processes and More Traditional Ones

At first, AM processes essentially allowed us to make plastic parts. Therefore, one of the first studies leading to the comparison of AM with other processes was interested in plastic injection [12, 13]. In these studies, Telenko et al. compared both processes from an electric energy consumption point of view. There is a large discrepancy between monetary and energy crossover volumes; this indicates that SLS may be more cost effective than energy efficient in some cases. In fact, the results of this comparative analysis of SLS and injection moulding indicate that manufacturers can save energy using SLS for parts with small production volumes.

Fig. 5 Part for process comparison [14]

Fig. 6 Breakeven analysis
comparing conventional
high-pressure die-casting and
selective maser sintering [14]

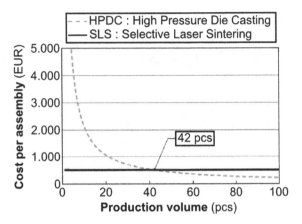

Energy crossover production volumes are much larger for a small part, indicating that specific crossover production volumes are sensitive to the size and geometry of the part to be produced. Nevertheless, this study does not take into account the manufacturing of the mould. This should be completed to integrate all the necessary data for an environmental analysis.

Atzeni et al. evaluated the production volume for which AM processes (selective laser sintering) are competitive with respect to conventional processes (high-pressure die-casting) [14]. In this study, they took into account the possibilities offered by AM (less material, less assembly; Fig. 5). Using an example of an aircraft part, they concluded that the breakeven point is estimated for a production of 42 components made of aluminium alloy as shown in Fig. 6.

In their study, Ruffo et al. compared, with the same criterion, injection moulding and AM processes (Fig. 7) [15]. Figure 8 shows different breakeven points between injection moulding and AM techniques for the different cost models utilised, with a comparison to the Hopkinson and Dickens model [16]. The breakeven point moved from 8000 to 14,000 parts, for plastic materials.

Fig. 7 Lever, object of the
study [15]

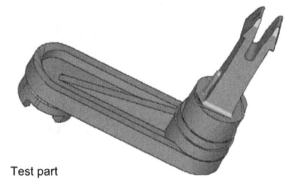

Test part

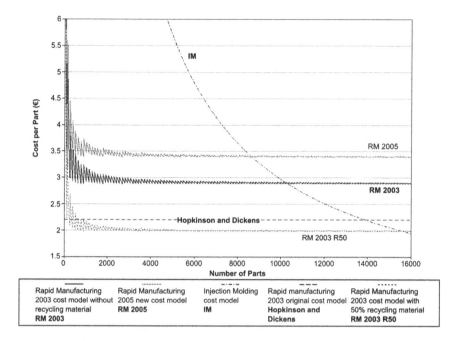

Rapid Manufacturing 2003 cost model without recycling material **RM 2003**	Rapid Manufacturing 2005 new cost model **RM 2005**	Injection Molding cost model **IM**	Rapid manufacturing 2003 original cost model **Hopkinson and Dickens**	Rapid Manufacturing 2003 cost model with 50% recycling material **RM 2003 R50**

Fig. 8 Cost model comparison [15]

With the same process comparison purpose, Morrow et al. provided a study to compare AM with machining [17]. The case studies were on a mould insert and a mirror and revealed that the relative energy consumption of machining versus AM is driven by the solid-to-cavity volume ratio. At low ratios, an AM pathway minimises energy consumption and emissions, whereas at high ratios the CNC milling pathway minimises energy consumption and emissions.

More recently, Serres et al. proposed a study comparing an AM process (CLAD, a directed energy deposition process) and machining on a mechanical part manufactured out of titanium alloy (Fig. 9) [18]. This study helps to highlight that on the

Fig. 9 Test part for the study [18]

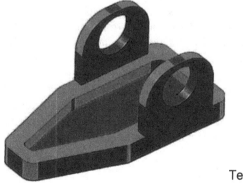

Test part

Fig. 10 Environmental
impact assessment of the test
part, considering two
processes [18]

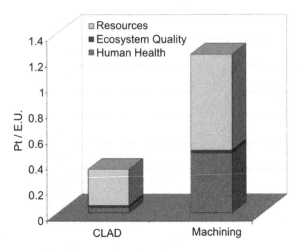

whole life cycle, from raw material extraction to manufacturing, AM reduces about
80 % of the environmental impacts (Fig. 10). Nevertheless, this study took into
account only one part geometry for both processes, and did not consider design for
manufacturing rules for optimising geometries from a manufacturing point of view.

Recently, Faludi et al. compared additive manufacturing versus traditional
machining via life-cycle assessment [19] and Yoon et al. did a comparison of
energy consumption in bulk forming, subtractive and additive processes [20]. They
characterised the processes via their specific energy consumption (SEC), in J mm^{-3}
or KWh kg^{-1}. The values of the SEC of similar additive manufacturing processes
are so different, with lots of uncertainty on the method of calculation, that it is
practically impossible to use SEC for an environmental performance assessment.

2.2.5 Considering Energy Consumption and Quality of the Part

Of course, a part whose geometric quality does not meet the specifications will not
be accepted even if the electric energy consumption during production has been
minimised.

Strano et al. have studied the correlation between the final surface roughness of
the part produced and the energy consumption of the machine [21]. This study
investigated a computational technology for the identification of optimal part orientations for the minimisation of surface roughness and simultaneously energy
consumption in the manufacturing process. Figure 11 shows the sample geometry
to be manufactured and Fig. 12 the related optimisation, represented by the Pareto
set. The results show that, moving along the Pareto front, although most solutions
have similar values of energy required to manufacture the part, choosing certain
angles allows part quality to be increased considerably.

Fig. 11 Artefact to be manufactured [21]

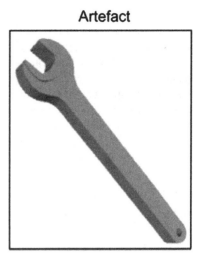

Fig. 12 Related pareto solutions [21]

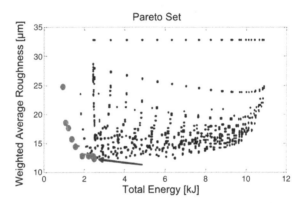

This study was partially based on the modelling of the surface roughness previously proposed by Campbell et al. [22] and was recently completed in another study by Strano et al. [23].

2.2.6 Synthesis of Electric Energy Consumption Studies

This section of the literature review refers to a number of studies on electric energy consumption in additive manufacturing processes. It can be seen that it is important to consider the set part process when characterising such a process. The morphology of the part produced as well as its position and orientation in the machine space have strong influences on the final results.

Table 1 summarises all studies concerned with electric energy consumption. The specific energy consumption (SEC, in KWh/kg) is used to compare the different processes.

This table allows a first comparison between AM processes. In this table, five technologies have been studied. It is still difficult to make a machine choice, considering which one has the least environmental impact, because these machines do not allow the production of parts with identical specifications. For example, stereolithography will produce prototypes whose lifetime is limited, unlike selective laser melting or electron beam melting which will realise functional parts, whose lifetime may well be longer.

Table 1 Comparison of specific energy consumption

Technology	Machines	Materials	SEC (KWh/kg)	Parts number[a]	References
Stereolithography	SLA-250	Epoxy resin SLA 5170	33	c	[5]
	SLA-3000	Epoxy resin SLA 5170	41	c	
	SLA-5000	Epoxy resin SLA 5170	21	–	
Selective laser sintering	Sinterstation DTM 2000	Polyamide	40	c	
	Sinterstation DTM 2500	Polyamide	30	c	
	Vanguard HiQ	Polyamide	15	b	[24]
	EOSINT M250 Xtended	Metallic powder (Bronze + Ni)	710	1	[8]
	EOSINT P760	Polyamide PA2200 balance 1.0	37	63	[25]
		Polyamide PA2200 speed 1.0	40	12	
		Polyamide PA3200GF	26	11	
Fused deposition modelling	FDM 1650	ABS plastic	346	c	[5]
	FDM 2000	ABS plastic	116	c	
	FDM 3000	ABS plastic	697	1	[8]
	FDM 8000	ABS plastic	23	c	[5]
	FDM Quantum	ABS	202	c	
Selective laser melting	MTT SLM 250	Metallic powder SAE 316L	31	6	[10]
Electron beam melting	Arcam A1	Metallic powder Ti-6Al-4 V	17	5	

[a]Number of parts built in the same time during the experiments
[b]Fabrication of the entire build volume of the machine ($380 \times 330 \times 340$ mm^3)
[c]Calculation depends on the material flow

This table shows the environmental impact of the manufacturing phase, due to electric energy consumption. However, for a more complete environmental assessment, material consumption also has to be taken into account. That is the main point of the next section.

2.3 Raw Material Consumption

2.3.1 Introduction

Additive processes are seen as environmentally interesting because they seem to consume only the required material for the production of the final part. Nevertheless, whatever the technology, it cannot be considered that all the raw material consumed is found on the final part.

In 3D printing, it is necessary to consider material consumption to create the supports needed to manufacture the part. These supports will be subsequently removed either by dissolving or manually removing them. Similarly, when using selective laser melting technology, an amount of the powder present in the workspace may not be reused [26]. In powder bed or powder projection technologies, a part of the deposed material is not fused, and it is necessary to consider this raw material lost in the environmental analysis. In powder bed, all the powder present in the workspace is not merged, fused, or sintered and could require a post manufacturing treatment to be reused.

2.3.2 Powder Recycling

The use of plastic (and, of course, metallic) powders requires some attention. In fact, plastic powders are sensitive to aging which reduces their mechanical properties [27].

To avoid premature aging of plastic powders, Dotchev et al. have developed a methodology to recycle the unsintered powders [26]. In this study, they analysed the influence of the recycled powder rate mixed with fresh powder on the final part quality. The objective was to limit the 'orange peel' texture on the parts produced. Finally, they defined a methodology that could improve the powder quality control, minimise the part quality variation, and reduce the amount of fresh powder used in the laser sintering process.

Metallic powders may be sensitive to the moisture contained in the air, causing their oxidation. Usually, the nonfused powder is reused after sieving treatment, and a few studies are focused on recycling the metallic powder in AM processes.

2.4   Other Flows that Affect the Environment

The environmental performance assessment of a manufacturing process must necessarily take into account all of the flows through the process (input and output). Even if AM processes use less consumables than most conventional ones, it is therefore not possible to assess the environmental performance by considering only the electric energy consumption. The quantity of raw material used as well as waste produced during the process, all the fluids such as inert gas to prevent oxidation, and cooling fluids for the machine must also be taken into account because they contribute to the overall environmental impact.

Kellens et al. developed the UPLCI (unit process life-cycle inventory) methodology for systematic analysis of the manufacturing process [28]. They applied it to the selective laser sintering process [29–31]. This methodology took into account all the flows through the system. They analysed the electric energy consumption, compressed air consumption, and material consumption, and took into consideration the environmental impact due to powder, consumables, and emission. Figure 13 shows a schematic overview of the parametric estimation model for the SLS process.

The methodology developed by Kellens et al. allows us to comment, analyse, and improve the process knowledge, especially for manufacturing prototypes or small batch sizes. The knowledge generated by this methodology allows us to bring data on manufacturing processes to LCA databases [32]. In their work, the UPLCI methodology was applied on additive manufacturing, laser cutting, and EDM. Finally, it proposed new ways to improve these processes from an environmental

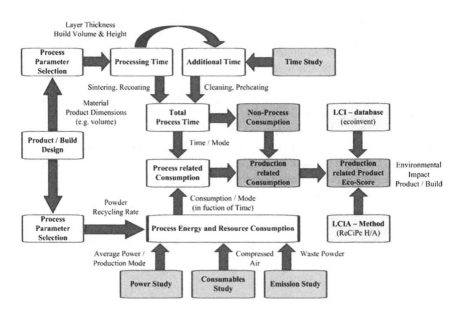

Fig. 13 Overview of the parametric impact estimation model for the SLS process [31]

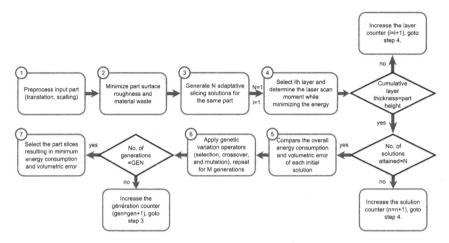

Fig. 14 The developed optimisation framework for adaptive slicing [9]

point of view, based on the electric energy consumption and on the material consumption, but also on the architecture of the machines.

Similarly, Verma et al. offered a study, also cited in Sect. 2.2.2, oriented not only on the optimisation of energy consumption but also focused on material consumption [9]. Considering these two consumption factors, they set up a multiobjective optimisation to minimise overall material consumption and power consumption. Furthermore, they imposed a certain quality of the part by coupling the aforementioned minimisation of consumption with maximising the surface quality (controlled by the surface roughness). In this study, they set up a double-loop optimisation. Initially, they optimised the overall part, minimising the amount of raw material and surface roughness. Secondly, they optimised, for every slice of the part, the electric energy consumption. Figure 14 summarises the optimisation algorithm.

This consideration of all consumption flow is an essential step for the characterisation of manufacturing processes. Studies taking into account these remarks are very recent and need to be developed with further investigations.

2.5 Possibilities Offered by AM Processes on the Whole Life Cycle of a Product

The studies presented in the previous sections are generally centred on the process. These studies help to compare the manufacturing processes between themselves, allowing us to make a choice on the most environmentally friendly technology in the manufacturing stage. But even if the environmental impact due to the manufacturing phase may be important, it may be negligible when considering the whole life cycle of the part. On this point, additive manufacturing may offer interesting design of parts, from an environmental point of view, on the whole life cycle. In this case, analysing the possibilities offered by the additive manufacturing process, such as topology optimisation, optimised design minimising mass, multifunction integration, and the like, could help designers to create an additive manufactured part with fewer environmental impacts than a machined one.

The collaborative project Atkins was interested in this issue [33, 34]. In this project, the authors studied all the possibilities for reducing the environmental impact of parts produced by AM processes. Apart from the already mentioned advantages in design, AM also reduces the availability time and the impacts generated by transportation (from production stage place to use stage place). Indeed, manufacturing facilities can be built close to the use stage location. The part to be produced is sent as a numerical file and will be realised close to the place of consumption. Then, this reduces the environmental impact caused by the transportation stage, which is a source of significant environmental impacts. Manufacturing companies also take advantages from additive manufacturing because they need very little time to adapt their production chain at the market; the changeover time is considerably reduced.

The Atkins project helped to highlight the possibilities of AM in order to minimise the overall environmental impact of a product. Moreover, as a result of the project, a software tool was developed as a guide in the choice of processes, with purposes to minimise the environmental impact or the economic impact. One of the major conclusions of this project was that additive manufacturing can be of great benefit in the aeronautic and transport fields, because of the mass minimisation opportunities for embedded parts.

2.6   Synthesis of the Literature Review

Efforts to characterise the environmental performance of AM processes have often focused on electric energy consumption. In this section different studies were analysed. One can realise that only a few studies were concerned with the raw material consumption or fluid consumption for these processes.

This lack of data is probably due to the youth of AM processes. However, as has already been noted in this section, it is important, in the environmental analysis context, to take into account all the flows through the process in order to assess its environmental performance precisely.

3   Environmental Impact Assessment Methodology

3.1   Introduction

In the third part of this chapter, a methodology to assess the environmental impacts of an additive manufacturing process is presented.

In the next section, as a general approach of the process leading to the production of a mechanical part, all the life-cycle stages of the part (from raw material to end of life) must be taken into consideration to correctly evaluate the environmental impacts.

In a third section, the manufacturing process is the main point. AM process modeling from an environmental point of view is done. The environmental impacts generated at this stage are mainly due to the resource consumption (material, electric, etc.) and waste production (support, etc.). An approach coupling all consumptions is presented.

Finally, the fourth section summarises the contributions of such a methodology.

3.2 General Approach

The methodology for evaluating the environmental impacts of AM processes that is presented in this chapter aims to open the scientific locks that have been outlined in the literature review.

This methodology, based on an accurate knowledge of manufacturing processes, allows the analysis of the environmental performance and takes into account two aspects. The first one is interested in the whole life-cycle stages of the part (Fig. 15). The second is focused on the process's objectives to estimate quantitatively all the resource consumption of the set part process (Fig. 16).

The production of mechanical products is generally made by the succession of stages. Indeed, parts are rarely produced directly by using only one single process. Figure 15 shows one of these sequencing stages.

It is thus necessary to take into account all the stages needed for the manufacturing of the part. Indeed, a vision being interested only in one stage can lead to a wrong analysis because the environmental impacts which could be minimised during a stage can be drastically increased in another one. For instance, when

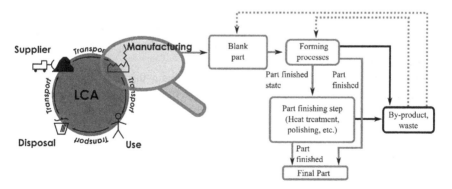

Fig. 15 Life-cycle stages and the manufacturing phase

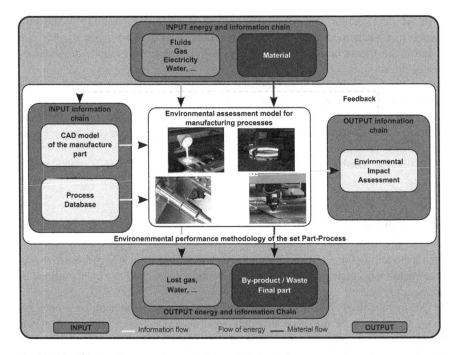

Fig. 16 Methodology for assessing environmental impact of a set part process

considering a directed energy deposition process such as projection of powders, it seems little sensible to be interested only in the manufacturing stage without being concerned in the stage of production of powder or the finishing postprocess. That is why a global approach, as proposed by life-cycle analysis, must be used to estimate the environmental performance of a manufacturing process.

Then the proposed methodology is thus interested in all the stages of manufacturing a product. In particular, the stages taken into account are listed below.

- *Raw material production.* In an AM process, raw materials are most of time plastic filament, plastic or metallic powder, or liquid resin.
- *Shaping.* This stage consists, thanks to a set of manufacturing substages, in transforming the raw material, obtained in the previous stage, into a finished or semi-finished part.
- *Postprocessing.* This last stage allows us to obtain the final dimensions and expected characteristics of the part. For example, it concerns the support removal operations; some finishing processes such as machining, polishing, or laser polishing; and so on.
- *Waste recycling.* This waste can be of various materials such as unsintered powder or support material, or fluids needed during manufacturing, among others.

The knowledge of all manufacturing stages is essential to propose a global and accurate assessment of environmental impacts. From this knowledge, it is then possible to define a modelling of each stage from an environmental point of view. This is the main topic of the next section.

3.3 Manufacturing Process Modeling

3.3.1 Framework and Limits

In life-cycle analysis, data concerning the manufacturing processes could be extracted from databases (for instance, Ecoinvent). What is found in these databases is just a macroscopic vision of the processes, with a global average value for characterising each process. Now this global vision does not allow us to take into account the influence of the manufacturing parameters (strategies, feed rate, temperature, etc.) on the final energy consumption. So these data often suffer from accuracy. Furthermore, they are only a 'picture' of a process and do not allow us to put forward the relation between manufacturing parameters, the part to be produced, and the total environmental impact. And AM processes are still not referenced in such databases.

Therefore, a modelling of the manufacturing stages, taking into account all input parameters of the machine is necessary. The objective of this model is at first to be able to predict all the consumption-generating impact during the process, and then, secondly, to set up a minimisation loop by modifying design parameters or manufacturing parameters.

Figure 16 presents the global vision of the developed methodology. The figure illustrates the necessity of taking into account all the flows of materials, energies, and information in the manufacturing stages modelling.

From the well-detailed knowledge of the manufacturing process, translated as predictive models, it is then possible to link part design and environmental impacts during manufacturing. The aim is thus to link the environmental impact due to the part production to its numerical model.

The first stage necessary is to define all the flows which will be taken into account and consequently the limits of the study. Indeed, in the developed methodology, even if it has been suggested to take all the flows, it is quite evident that certain limits must be set before completing the study.

Figure 17 shows the limits imposed on a manufacturing process, in the case of a directed energy deposition process. Similar limits may be easily constructed for other processes. In this figure, it can be noticed that the consumption of inert gas, compressed air, hydraulic fluids, and metallic powders, as well as electricity are taken into account during the environmental impact assessment. It is important to underline that the chosen limits are similar to the 'system boundaries' as defined in the standard ISO 14955-1, Machine tools—Environmental evaluation of machine tools, Part 1: Design methodology for energy-efficient machine tools [35]. Indeed,

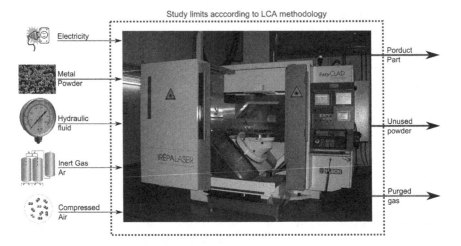

Fig. 17 Directed energy deposition (CLAD) process

the inert gas production, hydraulic fluids production, and compressed air production are not included in the system boundaries. The manufacturing of the machine is also not within the scope of the study. It would be possible to extrapolate the study by including the production of all the inputs and the manufacturing of the machine. It would be interesting because such a study would allow us to show that an optimal choice of components (axes motor drives, for instance) as well as the architecture of the machine allows the optimisation of its energy consumption during its use phase (when manufacturing a product). These studies have already been conducted by Kroll et al. [36] and Nuyen et al. [37].

3.3.2 Input Data

In the methodology, the goal is to remain centred on the set part process. Indeed, based on the literature review in the previous section, geometry of the part as well as its positioning in the machine workspace could influence the process consumption in terms of material or energy.

The major input data of the methodology is the numerical model (CAD model) of the part. It could allow the modification of its geometry, and furthermore, it is possible to advise designers with a software tool which will indicate the areas of the part for which the environmental impact could be optimised.

The second input data are based on a well-detailed knowledge of the manufacturing process, more specially the process parameters, path trajectories, axes motor drives, cooling unit system, and the like. This knowledge is stored in a database, defining the AM process in the set part process, which will be used during the environmental impact assessment.

3.3.3 A Multistep Methodology

Figure 18 presents a global view of the developed methodology. This methodology has for its objective to link the environmental impact (output) to the numerical model (input) in a set part process approach.

Different methods can be used to classify the impacts caused on the environment. In this study, the method is Eco-Indicator 99, which is a method-oriented damage classifier and translates all the impacts into a unique point value, a nondimensional number used to compare the different sources of impacts [38]. The value of 1 point is defined by a thousandth of the environmental impact caused by a common European during a year. For comparison, the production of 1 kg of primary steel is around 100 mPts and the production of 1 kg of stainless steel is around 900 mPts. The choice of this method has been made because it was the one used in most of the studies on environmental assessment of additive processes that have been analysed in the state of the art.

This methodology is decomposed into four steps:

- Numerical programme generation
- Extraction of the command parameters

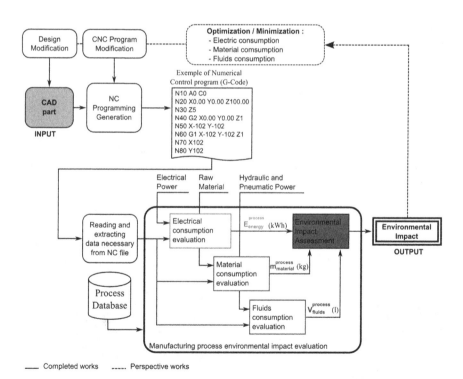

Fig. 18 Environmental performance assessment methodology of the set part process

- Construction of process database
- Environmental impact assessment

These steps are detailed in the next section, which also gives complementary information on the methodology.

4 Application to Directed Energy Deposition

4.1 Introduction to Directed Energy Deposition Process

This study is based on a directed energy deposition process, known as the CLAD process, which manufactures 3D metallic parts from a CAD model. In this process, a five-axis deposition nozzle, where metallic powders are injected into the laser beam, creates a small melt pool on the workpiece which is cooled down when the laser beam moves on. The part is built as the nozzle moves. Figure 19 shows the design of the nozzle, with laser beam, and an example of a part produced by the machine. The machine is equipped with two kinds of nozzles, which allows us to obtain a welding bed from 0.8 mm (the MesoCLAD nozzle) to 4 mm (the MacroCLAD nozzle). The machine structure is a five-axis machine tool (Huron KX8), with its conventional machining spindle (for machining operations such as finishing), in which were added the two nozzles, two powder feeders (for raw material powder), and a 4-kW fibre laser.

4.2 Atomisation of Raw Material

The first step for the process is to produce powder (metallic, ceramic, glass) which will be introduced in the machine. An atomisation process is used to obtain this powder (Fig. 20).

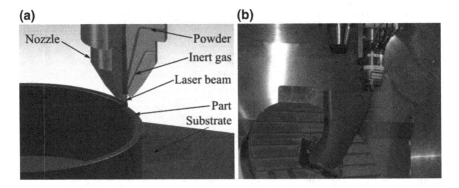

Fig. 19 a CLAD nozzle design; **b** example of part produced by this AM process

Fig. 20 Atomisation process

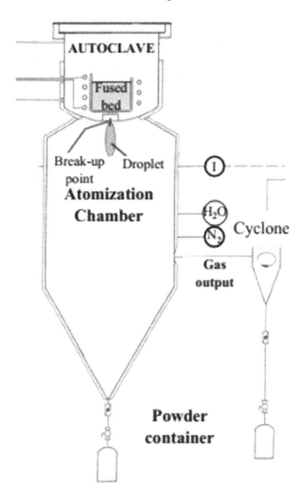

In this process, raw materials (from block or cylinder) are heated to melting point in a chamber and then atomised with an inert gas (in the case study: argon). This atomisation consists of compressing, under high depression, the metallic fluid which will be atomised into small droplets in reaction to depression.

In this process, many values can be saved and it is possible to establish a model for the atomisation step. The model is made with experimental values such as

- Gas consumption
- Water consumption
- Electrical consumption

Table 2 shows all the parameters that have been monitored or calculated during experiments on the atomisation process in order to build the modeling of the process.

Table 2 Nomenclature for atomisation process

Parameters	Name	Units	Saved/calculated
V_{argon}	Volume of consumed argon	Cubic metre	Calculated
d_{argon}	Argon flow rate	Cubic metre per second	Monitored
ρ	Gas density	Kilogram per litre	–
V_{water}	Volume of consumed water	Litre	Calculated
d_{water}	Water flow rate	Litre per second	Monitored
$t_{atomization}$	Time for atomisation	Second	Monitored
$E_{electrical}$	Electric energy	KWh	Calculated
$P_{inductor}$	Electrical power of the inductor	Watt	Monitored
$P_{depression}$	Electrical power of the vacuum system	Watt	Monitored
t_{vacuum}	Total time of the vacuum system ON	Second	Monitored
$P_{preheating}$	Electrical power of the preheating system	Watt	Monitored
$t_{preheating}$	Total time of the preheating system ON	Second	Monitored

4.2.1 Gas Consumption

Gas consumption is linked to the volume of the inert chamber and the atomisation step. Figure 21 shows the variation flow of argon in the chamber, during the atomisation of 1 kg of metallic glass.

From these experimental data, an empirical modelling for gas consumption is determined, according to Eq. (1):

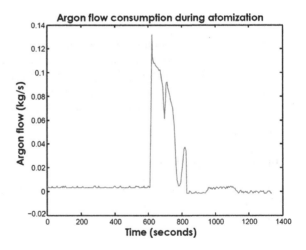

Fig. 21 Argon flow consumption

$$V_{argon} = \frac{1}{\rho} * \int_0^{t_{atomisation}} d_{argon} * dt \qquad (1)$$

4.2.2 Water Consumption

In this system, water runs in a closed-loop system. However, an amount of used water is released into nature and a corresponding amount of fresh water is obtained because the cooling system is not efficient enough. The total volume of consumed water is calculated according to Eq. (2):

$$V_{water} = d_{water} * t_{atomisation} \qquad (2)$$

4.2.3 Electrical Consumption

Electrical consumption is due to different features of the machine (inductor, pre-heater, vacuum pump). Figure 22 shows a profile of the inductor electrical consumption during the atomisation process.

From this experimental monitored value, an empirical model for electrical consumption is determined according to Eq. (3).

$$E_{electrical} = P_{depression} \times (t_{atomisation} + t_{vacuum}) + P_{preheating}$$

$$\times t_{preheating} + \int_0^{t_{atomisation}} P_{inductor} \cdot dt \qquad (3)$$

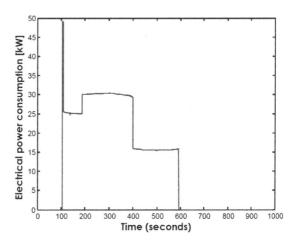

Fig. 22 Inductor electrical power consumption during atomisation

Table 3 Values of the experimental data monitored and calculated

Input consumption	Value
Gas consumption	7 m^3
Water consumption	155 l
Electrical consumption	4 kWh
Efficiency	46 %

Table 3 presents the results of the study for 1 kg of glass powder atomisation. These values will help to elaborate the complete environmental assessment.

4.3 Environmental Performance Modeling for the AM Process

According to the methodology presented in Fig. 18 the environmental impacts generated in the manufacturing stage are modeled from three inputs:

- Electrical consumption
- Material consumption
- Fluids consumption

For each input's consumption, a model based on an empiric model or analytical model has been developed. These models allow evaluation of the global environmental impact of the part from its CAD model. From the CAD model, a G-code file is created which will give the instruction for the machine. From this file, every parameter required to evaluate the environmental impact is extracted.

As well as what has been done for the atomisation process, Table 4 shows all the parameters monitored during the experiments or calculated.

For environmental impact assessment, the Eco-Indicator 99 has been used, with the following characterisation factors:

- $fc_{\text{argon}} = 1.78 \text{ mPts.kg}^{-1}$;
- $fc_{\text{material}} = 86 \text{ mPts.kg}^{-1}$;
- $fc_{\text{elec}} = 12 \text{ mPts.kWh}$, corresponding to the French electricity production characterisation factor.

4.3.1 Fluid Consumption

Fluid consumption is due to the inert gas used during the process which allows us to project and protect metal powder in the melting pool. In this study, the inert gas is argon; it is the same gas for the two functions. Its consumption varied during the manufacturing step and depends on the part morphology. An environmental impact is associated with the inert gas consumption during the manufacturing step, according to Eq. (4).

Table 4 Nomenclature for directed energy deposition process

Parameters	Name	Units	Saved/calculated/machine knowledge
EI_i	Environmental impact for substance i	mPts	Calculated
t_{man}	Manufacturing time	Second	Monitored
d_c	Desired carrying gas	Kilogram per second	Monitored
d_f	Desired forming gas	Kilogram per second	Monitored
k	Weight factor (lost/fused powder)	–	Machine knowledge
d_p	Powder flow rate	Kilogram per second	Monitored
e_n	Nozzle efficiency	Kwh	Machine knowledge
$g(Pl)$	Function for laser electrical power consumption	–	Monitored
t_{laser}	Switch-on time such as $t_{man} = t_{laser} + \overline{t_{laser}}$	Second	Monitored
$Pc_{stand\text{-}by}$	Power consumed by the cooling system in standby mode	Watt	Monitored
Pc_{on}	Power consumed when the cooling system works	Watt	Monitored
Pe_i	Electrical power consumed by the i-axis	Watt	Monitored
Pe_{idle}	Constant electrical power demand	Watt	Monitored

$$E.I._{\text{fluids}} = [dc + df] * t_{man} * fc_{argon} \tag{4}$$

4.3.2 Material Consumption

Now, the focus is put on the determination of the powder consumption during part manufacturing. In fact, an advantage of the additive manufacturing process is to project and fuse exclusively the necessary powder. However, this is not the reality and an amount of powder will not be fused in the directed energy deposition process.

In the studied machine, two different kinds of nozzles can be used to project the powder. Their efficiency is not the same. Moreover, the efficiency of each nozzle depends on the desired powder flow rate.

An analytic model is proposed for the material consumption estimation during part manufacturing, according to Eq. (5):

$$EI_{material} = [en + k * (1 - en)] * dp * t_{man} * fc_{material} \tag{5}$$

4.3.3 Electric Consumption

In each machine, electric components can be classified into two categories. Some features have constant energy consumption such as electrical cabinet and hydraulics components. For the other components, their electrical energy consumption depends on the part design but also on machine parameters.

The modelling of each feature of the directed energy deposition machine has been done and published by Le Bourhis et al. [39]. In this section, the results are summarised with Eq. (6) in which can be found the environmental impact of each component.

$$E.I_{\cdot electricity} = \left(g(Pl) * t_{laser} + Pc_{standby} * t_{man} + (Pc_{on} - Pc_{standby}) * t_{on} + \left(\sum_{i=1}^{5} \int_{0}^{tman} Pei(t) * dt \right) + Pe_{idle} \right) * fc_{elec} \tag{6}$$

4.3.4 Lost Powder Recycling

In this process, a nonnegligible amount of material is projected but not fused. It seems important to propose a method to recycle this powder. In fact, AM processes could be seen as environmentally friendly only if all the powder projected is used.

The lost powder cannot be used without treatment. In fact, this powder could cause several damages to the machine and needs to be sieved and dried before being reused. Some studies have been conducted to determine that this recycled powder has the same mechanical properties as fresh powder.

4.4 Industrial Example

4.4.1 Case Study Introduction

The example below illustrates the possibility of the environmental impact assessment methodology. It is based on a case study presented by Le Bourhis et al. [40].

This example is an aeronautic part which is at this time produced by conventional machining. More than 80 % of raw material is machined to produce this part. In this example, the focus is on nozzle choice. As it has been previously mentioned, this directed energy deposition process uses two kinds of nozzle. Which one is more 'environmentally friendly'? The methodology helps to answer this question.

4.4.2 CAD Part

The part presented (Fig. 23) is composed, amongst others, with a pocket 200 mm square and 80 mm in depth. The part thickness is 4 mm. In this study, the answer is how to know which nozzle is better to manufacture the pocket. In fact, is it possible to choose in the NC programme generation which nozzle will be used?

4.4.3 Different Manufacturing Strategies

In this case, with the nozzle called MacroCLAD, the part can be produced in one trajectory by layer but the laser power demand will be very high (around 3 kW). However, if the smaller nozzle, called MesoCLAD, is used, the part needs five trajectories of 0.8 mm width by layer with a smaller laser power demand (around 250 W). The methodology developed allows us to choose which nozzle must be used to minimise the environmental impact of the manufacturing process.

4.4.4 Environmental Impact Results

The model used enables the evaluation of the environmental impact of each manufacturing strategy. This methodology is formalised on an informatics tool for designers. The first step is to read the G-code of the CAD model and extract all the values that are needed to evaluate the environmental impact such as laser power, trajectories, axis speed, and so on. From these values it is possible to calculate and preprocess the expected consumption. The results are given either in scientific units (kWh, litre, or kilogram) or in environmental units (mPts). The second unit allows comparing the different flow consumptions amongst them.

The results are shown in Fig. 24 and Table 5.

Fig. 23 Part model

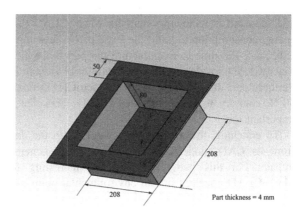

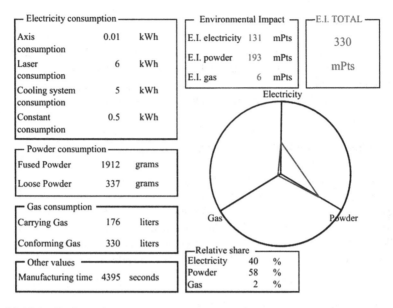

Fig. 24 MacroCLAD results

Table 5 Complete study results

Input consumption	Scientific units		Environmental impact (mPts)	
	MacroCLAD	MesoCLAD	MacroCLAD	MesoCLAD
Electricity	12 kWh	109 kWh	131	1332
Powder	2.249 kg	3.824 kg	193	328
Fluids	0.5 m^3	9.5 m^3	6	122
Time	4395 s	78,872 s		

These results show two different kinds of consumption. In fact, even if the power laser demand is more important for MacroCLAD than for MesoCLAD, the total energy consumption to build the same part is less important for MacroCLAD. That is because the time to manufacture the part is drastically reduced when using the MacroCLAD nozzle (in this case study, it obviously depends on the CAD model). Furthermore, the efficiency of the MacroCLAD nozzle is more efficient, around 80 % contrary to 35 % for MesoCLAD. Thus the powder consumption is less important too.

To conclude, the methodology would help the designer to determine, directly from its CAD model, which process could generate the lesser environmental impact. For this part, it should be interesting to manufacture it with the MacroCLAD nozzle, from an environmental point of view.

5 Conclusion

In this chapter, two main points are developed. First, the literature on all the studies led to characterising the environmental impact of AM processes. The studies on electric energy consumption of these processes are analysed, and then some studies taking into account raw material and all the flows through the process are detailed.

Second, a new methodology in order to evaluate, with accuracy, the environmental impact of a part from its CAD model is presented. In this methodology, the work is not focused only on electrical consumption but also on fluids and material consumption which also contribute to the environmental impact. In addition, the inputs of this methodology correspond to the set part process, which allows taking into account different manufacturing strategies and their influences on the global environmental impact. The methodology developed is based on both analytic models (validated by experiments) and experimental models.

And finally, an industrial example shows that for some manufacturing strategies, the environmental impact due to electrical consumption is not the predominant one. In this case study, material consumption has an important impact and has to be taken into consideration for a complete environmental impact assessment.

References

1. Kruth J, Leu M, Nakagawa T (1998) Progress in additive manufacturing and rapid prototyping. CIRP Ann Manuf Technol 47(2):525–540
2. Bourell DL, Leu MC, Rosen DW (2009) Roadmap for additive manufacturing: identifying the future of freeform processing. The University of Texas at Austin, Austin
3. Hao L, Raymond D, Strano G, Dadbakhsh S (2010) Enhancing the sustainability of additive manufacturing. In ICRM2010—green manufacturing, pp 390–395
4. Luo Y, Ji Z, Leu MC, Caudill R (1999) Environmental performance analysis of solid freeform fabrication processes. In: International conference on electronics and the environment, pp 1–6
5. Luo Y, Leu MC, Ji Z (1999) Assessment of environmental performance of rapid prototyping and rapid tooling processes. In: Solid freeform fabrication symposium, pp 783–792
6. Sreenivasan R, Bourell DL (2009) Sustainability study in selective laser sintering—an energy perspective. In: Solid freeform fabrication symposium, pp 257–265
7. Mognol P, Perry N, Lepicart D (2005) Environment aspect of rapid prototyping: process energy consumption. In: 12th CIRP life cycle engineering, 2005
8. Mognol P, Lepicart D, Perry N (2006) Rapid prototyping: energy and environment in the spotlight. Rapid Prototyp J 12(1):26–34
9. Verma A, Rai R (2013) Energy efficient modeling and optimization of additive manufacturing processes. In: Solid freeform fabrication symposium, pp 231–241
10. Baumers M, Tuck C, Hague R, Ashcroft I, Wildman R (2010) A comparative study of metallic additive manufacturing power consumption. In: Solid freeform fabrication symposium, pp 278–288
11. Baumers M, Tuck C, Wildman R, Ashcroft I, Hague R (2011) Energy inputs to additive manufacturing: does capacity utilization matter? In: Solid freeform fabrication symposium, pp 30–40

12. Telenko C, Seepersad CC (1997) A comparative evaluation of energy consumption of selective laser sintering and injection molding of nylon parts. In: Solid freeform fabrication symposium, pp 41–54
13. Telenko C, Seepersad CC (2010) Assessing energy requirements and material flows of selective laser sintering of Nylon parts. In: Solid freeform fabrication symposium, pp 289–297
14. Atzeni E, Salmi A (2012) Economics of additive manufacturing for end-usable metal parts. Int J Adv Manuf Technol 62:1147–1155
15. Ruffo M, Tuck C, Hague R (2006) Cost estimation for rapid manufacturing—laser sintering production for low to medium volumes. Proc Inst Mech Eng Part B J Eng Manuf 220 (9):1417–1427
16. Hopkinson N, Dickens P (2003) Analysis of rapid manufacturing—using layer manufacturing processes for production. J Mech Eng Sci 217:31–39
17. Morrow W, Qi H, Kim I, Mazumder J, Skerlos S (2007) Environmental aspects of laser-based and conventional tool and die manufacturing. J Clean Prod 15(10):932–943
18. Serres N, Tidu D, Sankare S, Hlawka F (2011) Environmental comparison of MESO-CLAD® process and conventional machining implementing life cycle assessment. J Clean Prod 19(9–10):1117–1124
19. Faludi J, Bayley C, Bhogal S, Iribarne M (2014) Comparing environmental impacts of additive manufacturing versus traditional machining via life-cycle assessment. Rapid Prototyping J 21(1):14–33
20. Yoon HS, Lee JY, Kim HS, Kim MS, Kim ES, Shin YJ, Chu WS, Ahn SH (2014) A comparison of energy consumption in bulk forming, subtractive, and additive processes: review and case study. Int J Precis Eng Manuf Technol 1(3):261–279
21. Strano G, Hao L, Evans KE, Everson RM (2010) Optimisation of quality and energy consumption for additive layer manufacturing processes. In: ICRM2010—green manufacturing, pp 364–369
22. Campbell RI, Martorelli M, Lee HS (2002) Surface roughness visualisation for rapid prototyping models. Comput Des 34:717–725
23. Strano G, Hao L, Everson RM, Evans KE (2013) Surface roughness analysis modelling and prediction in selective laser melting. J Mater Process Technol 213(4):589–597
24. Sreenivasan R, Goel A, Bourell DL (2010) Sustainability issues in laser-based additive manufacturing. Phys Procedia 5:81–90
25. Kellens K, Yasa E, Renaldi, Dewulf W, Kruth J, Duflou J (2011) Energy and resource efficiency of SLS/SLM processes. In: Solid freeform fabrication symposium, pp 1–16
26. Dotchev K, Yusoff W (2009) Recycling of polyamide 12 based powders in the laser sintering process. Rapid Prototyp J 15(3):192–203
27. Choren J, Gervasi V, Herman T, Kamara S, Mitchell J (2001) SLS powder life study. In: Solid freeform fabrication symposium, pp 39–45
28. Kellens K, Dewulf W, Overcash M, Hauschild MZ, Duflou JR (2011) Methodology for systematic analysis and improvement of manufacturing unit process life-cycle inventory (UPLCI)—CO2PE! initiative (cooperative effort on process emissions in manufacturing). Part 1: methodology description. Int J Life Cycle Assess 17(1):69–78
29. Kellens K, Yasa E, Dewulf W, Duflou JR (2010) Environmental assessment of selective laser melting and selective laser sintering. In: Going green—CARE Innovations. no. Section 4
30. Kellens K, Yasa E, Renaldi, Dewulf W, Kruth JP, Duflou JR (2011) Energy and resource efficiency of SLS/SLM processes. In: Solid freeform fabrication symposium, pp 1–16
31. Kellens K, Renaldi R, Dewulf W, Kruth J, Duflou JR (2014) Environmental impact modeling of selective laser sintering processes. Rapid Prototyp J 20(6):459–470
32. Kellens K (2013) Energy and resource efficient manufacturing—unit process analysis and optimisation. University of Leuven, KU Leuven
33. Hague R, Tuck C (2007) ATKINS: manufacturing a low carbon footprint—zero emission enterprise feasibility study. Loughborough University, Loughborough
34. Reeves P (2011) Does additive manufacturing really cost the earth—stimulating am adoption through economic and environmental sustainability. In: TCT

35. ISO 14955-1: Machine tools—Environmental evaluation of machine tools (2014) Part 1: design methodology for energy-efficient machine tools
36. Kroll L, Blau P, Wabner M, Frieß U, Eulitz J, Klärner M (2011) Lightweight components for energy-efficient machine tools. CIRP J Manuf Sci Technol 4(2):148–160
37. Nguyen T, Ai TAL, Museau M, Paris H (2014) Methodology for design for energy efficiency of production system. In: IDMME—virtual concept—improve—ingegrag conference
38. Goedkoop M, Spriensma R (1999) The eco-indicator 99 methodology
39. Le Bourhis F, Kerbrat O, Hascoet JY, Mognol P (2013) Sustainable manufacturing: evaluation and modeling of environmental impacts in additive manufacturing. Int J Adv Manuf Technol 69:1927–1939
40. Le Bourhis F, Kerbrat O, Dembinski L, Hascoet J, Mognol P (2014) Predictive model for environmental assessment in additive manufacturing process. In: 21st CIRP conference on life cycle engineering, pp 1–6

Sustainability Based on Biomimetic Design Models

Henrique A. Almeida and Eunice S.G. Oliveira

Abstract Ecodesign is an approach to designing products with special consideration for the environmental impacts of the product's life cycle. In a lifecycle assessment, the life cycle of a product is usually divided into procurement, manufacture, use, and disposal. Ecodesign is a growing responsibility and understanding of our ecological footprint on the planet. It is imperative to search for new productive solutions that are environmentally friendly and lead to a reduction in the consumption of materials and energy while maintaining the desired performance of the products. Bearing this in mind, additive manufacturing has the capability of producing components with the lowest amount of raw material needed and the highest geometrical complexity. This work aims to present novel bioinspired design methodologies for the production of additive manufacturing products with lower amounts of material and higher performance. The bioinspired design considers natural cellular-based concepts for the definition of novel product definitions.

Keywords Sustainable manufacturing · Additive manufacturing · Biomimetics · Bioinspiration · Cellular materials

1 Introduction

Nature has gone through evolution since life itself began on Earth and, during the process, living organisms have evolved well-adapted structures and materials over geological time through natural selection [42]. As evolution continued, nature

H.A. Almeida (✉) · E.S.G. Oliveira
School of Technology and Management, Polytechnic Institute of Leiria,
Leiria, Portugal
e-mail: henrique.almeida@ipleiria.pt

E.S.G. Oliveira
R&D Unit INESC Coimbra, Coimbra, Portugal

© Springer Science+Business Media Singapore 2016
S.S. Muthu and M.M. Savalani (eds.), *Handbook of Sustainability
in Additive Manufacturing*, Environmental Footprints and Eco-design
of Products and Processes, DOI 10.1007/978-981-10-0606-7_3

evolved in such a form that it has reached optimised biological forms, designs, functions, and concepts [51]. These optimised biological functions are present in all levels from a biological or dimensional point of view. For instance, on a biological level, one may be inspired from the organism, behavioural or ecosystem, whereas on the dimensional level, the scales of inspiration range from the macro to the nanoscale. Humans have looked at nature for answers and inspiration to solving problems throughout our existence. Biologically inspired design, adaptation, or derivation from nature is referred to as 'biomimetics' (which derives from the Greek word *biomimesis*), meaning mimicking biology or nature [9].

The field of biomimetics is highly interdisciplinary. It involves the understanding of biological functions, structures, processes, and principles of various living or nonliving things found in nature by biologists, physicists, chemists, and material scientists, and the design and fabrication of various materials and devices of commercial interest by engineers, material scientists, chemists, and others [9]. Nature has solved engineering problems such as self-healing abilities, environmental exposure tolerance and resistance, surface performance (e.g., hydrophobicity, self-cleaning, high and reversible adhesion), self-assembly, and energy conversion and conservation [9]. But in order to use nature's full potential of inspiration, one must understand some basic issues, such as nature's basic principles and the necessary steps on how to adopt biomimicry in our everyday life. Nature's basic principles are [7]:

- Nature runs on sunlight.
- Nature uses only the energy it needs.
- Nature fits form to function.
- Nature recycles everything.
- Nature rewards cooperation.
- Nature banks on diversity.
- Nature demands local expertise.
- Nature tends to minimise excesses from within.
- Nature taps the power of limits.

After understanding these principles, the necessary steps on how to implement biomimicry are [50]:

1. Identifying the challenge: Develop a design brief of human needs.
2. Interpreting the design briefs: Biologising the question: ask from nature's perspective.
3. Discovering natural models: Look for champions in nature who have solved the given challenge.
4. Abstracting the design principles: Find the repeating patterns and processes within nature that have achieved success.
5. Emulate nature's strategies: Develop ideas and solutions based on the natural model.
6. Evaluate against life's principle: Evaluate the design against life's principles.

Another aspect that must be considered, is that nature has managed to evolve and develop in a sustainable manner. Sustainable development is the development that meets the needs of the present without compromising the ability of future generations to meet their own needs [8, 32]. In other words, a balance between a society of any biological species and environment must be accomplished. With the coming of mankind, a third variable has emerged and brought imbalance to this system, namely the economic aspect [8, 32]. Based on this, there has been a growing concern from governments, companies, and the general public for a more sustainable development mainly due to an increase in the environmental and social impacts of human actions [8, 32]. The degradation of natural resources has been aggravated by several factors such as the actual lifestyle of the consumer societies, the rapid growth of emerging countries, inequalities between regions, and the reduction in the life cycle of each product requiring a high consumption of production and raw material resources [8, 32]. Sustainable industrial practices can contribute to the development of more sustainable products and processes. It is fundamental to apply a sustainable design framework in order to achieve and gain the triangular balance between society, environment, and economy. These principles are [36]:

1. Site and Climate Analysis: Analysing site, orientation, exposure, climate, topographical factors, local constraints, and natural resources.
2. Flexible Structural Systems: Investigating structural characters, permanence/temporariness, integration with building components, and so on.
3. Renewable Building Materials: Analysing efficiency of a material or a product, size, standardisation, structural adequacy, complexity, appropriateness, cost, labour involved, plantation origin, method of growth, embodied energy, recycled and reused content, toxicity, and the like.
4. Building Envelope Systems: Control of energy flows that enter (or leave) an enclosed volume, including consideration of orientation, seasonal variations, surrounding environment, function, and typology.
5. Modular Building Systems: Construction and assembling methods to facilitate substitution, repair, maintenance, diversified lifetime, and so on.
6. Renewable and Nonconventional Energy Systems: Integrating sources of energy that do not reduce or exhaust their point of origin.
7. Innovative Air Controlling Systems (heating, cooling, ventilating, and air conditioning): Implementing strategies to provide thermohygrometric and air quality comfort, exploiting mechanically regulated, hybrid, or, preferably, totally passive techniques.
8. Water Collection and Storage Systems: Adopting methods, systems, and strategies to collect, store, distribute, use, recycle, and reuse water.

Based on these design principles, among the existing manufacturing technologies, additive manufacturing is the most promising. Additive manufacturing is an innovative way of producing components and possessing good environmental characteristics [5, 21, 38]. Additive manufacturing has the potential of reducing resource and energy demands as well as process-related CO_2 emissions [3, 4, 14, 21, 35, 47]. According to Serres et al. [53], the energy consumed by additive manufacturing to produce parts is also inferior when compared to conventional machining processes [11, 12]. By utilising only the amount of material needed for the building of the final part, additive manufacturing technologies reduce the material mass and energy consumption when compared to conventional subtractive techniques by eliminating scrap, on top of eliminating the need for tooling and the use of environmentally damaging process enablers [21, 27, 55]. In fact, it is possible to obtain a functional part directly from a CAD model with only one manufacturing step, contrary to conventional processes which need several manufacturing steps or resources to produce a part, presenting both part and production flexibility [12, 33].

In addition to the issues mentioned above, additive manufacturing systems are capable of manufacturing parts without any design rules or methods, such as part redesign [31, 48], part customisation [13, 40], and industrial design [10]. From a designer's point of view, the advantage over conventional subtractive or formative technologies is well illustrated by the great design freedom, regarding shape, colour, and/or functionality. These design freedoms enabled by additive manufacturing technologies are described below [22, 62]:

- Shape complexity: It is possible to manufacture any shape and form allowing customised geometries, and also, shape optimisation is easily enabled and implemented in the physical model.
- Hierarchical complexity: These technologies enable the production of hierarchical multiscale structures from the microstructure through geometric mesostructure (0.1–10 mm) to the part-scale macrostructure. Basically, the parts may comprise features at one scale size, which may have other smaller features added to them, and each of those smaller features again may also have other smaller features added, during the same production phase.
- Material complexity: Different materials may be processed at one point or in one layer, at a time during production, enabling the manufacture of parts of multimaterial compositions with a designated property gradient. It is also possible to maintain the same material composition and vary the geometric/processing parameters while undergoing production, allowing the production of parts with specific designed property gradients.
- Functional complexity: When building parts in an additive manner, the inside of the part is always accessible. This makes it possible to produce self-assembled components, allowing the incorporation of multiple design solutions for multiple functionalities.

Considering the numerous design possibilities, and bearing in mind that one of the sustainable concerns regards raw material efficiency, it is obvious that a designer needs to become biologically inspired in order to produce optimum parts with all their required properties and functionalities while reducing the amount of raw material needed for production. This chapter presents both bioinspired lightweight material schemes, such as cellular structures, as well as engineered lightweight material schemes, for example, topology optimisation, for the production of additive manufactured parts.

2 Cellular Structures

Due to the referred properties found in nature, several research projects in different areas have emerged trying to mimic these materials and architectures or to develop mechanisms and technology which allow the manufacturing of bioinspired products [56]. Many natural materials (for instance, bone, wood, coral, and cork) are based on cellular designs, as they all have a porous construction optimising a specific physical property, such as strength, density, or stiffness. Their geometric structure has adapted to be multifunctional according to both biological and environmental conditions, for example, conveying fluids and nutrients or conducting heat [45, 49, 56]. Figures 1 and 2 present some existing natural cellular structures.

From a mechanical engineering point of view, a key advantage offered by cellular materials is the high strength accompanied by low mass. These materials provide good energy absorption characteristics and both good thermal and acoustic insulation properties. Cellular material structures, such as honeycombs and lattice structures, enable unprecedented stiffness and strength characteristics for a given weight [28, 49]. In order to achieve these optimal performance properties, cellular structures have, in general, very complex geometries that are prohibitive for the classical manufacturing processes. Based on the design freedoms provided by additive manufacturing, both the geometric complexity and scale size of natural cellular structures it is possible to fabricate them [45, 56].

Two typical types of cellular structures are the lightweight structures and compliant mechanisms. Lightweight structures are rigid and designed to reduce weight, while increasing strength and stiffness. Compliant mechanisms are designed to transform motions and forces [59]. The performance of lightweight structures can be enhanced by using adaptive cellular structures. Wang [59] proposes a new unit truss cell in order to design adaptive cellular structures. The unit truss approach facilitates the design of adaptive cellular structures for enhanced mechanical properties via geometric modelling, finite element analysis, geometry optimisation, and additive fabrication.

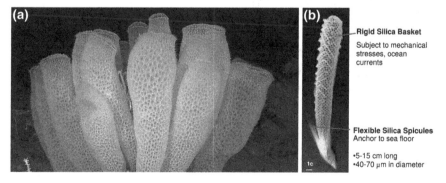

Fig. 1 Euplectella sponge, commonly known as Glass sponge **a** in its marine habitat; **b** showing the basket cage and the spicules (adapted from [42])

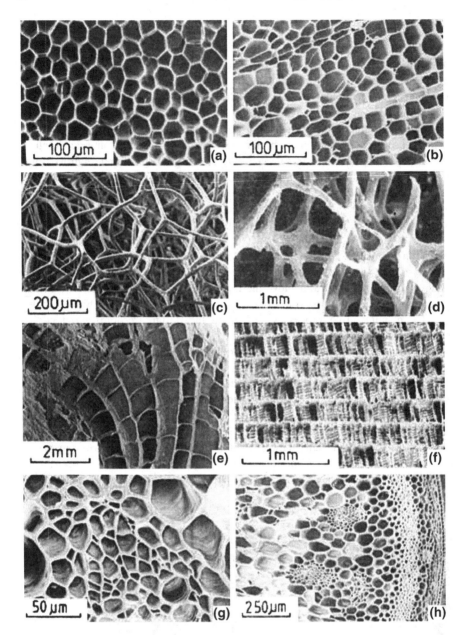

Fig. 2 Examples of cellular materials: **a** cork; **b** balsa; **c** sponge; **d** cancellous bone; **e** coral; **f** cuttlefish bone; **g** iris leaf; **h** stalk of plant [23–25]

Graded cellular structures are structures that can provide an adaptive and ordered porosity distribution and changing from thin struts with high porosity and small elasticity at one end of the part's boundary to thick struts with low porosity and large elasticity at the other end of the part's boundary. As mentioned before, additive manufacturing processes are capable of manufacturing these types of functionally graded structures. Wang et al. [60] used a graded cellular structure to design an uncemented acetabular prosthesis for enhanced stability on an implant–bone interface (Fig. 3). The proposed acetabular component demonstrates a better fatigue strength and significantly less wear. The metal-on-metal bearing replaces the metal-on-polyethylene bearing in this acetabular component.

Ziegler and Jaeger [64] present a method of generating cellular structure designs. In their design, the internal structure is based on a trabecular structure similar to the structure found in cancellous bone and similar to a structure proposed by Kowalcyk [34]. In this work a cantilever chair was chosen to demonstrate biomimetic structuring and mechanical optimisation (see Fig. 4). The results from the designed model are automatically used to optimise the mechanical strength of the trabeculae where needed. This mechanical optimisation can increase the diameter either in one direction (if the stresses are mainly along one direction) or in all directions (if the stresses are close to being hydrostatic). Two steps of strengthening the individual trabeculae have been chosen resulting in either a doubling of Young's modulus and strength or a fivefold increase in these values. Small bumps were added on the outer surfaces for aesthetic reasons, which are biomimetically inspired by Radiolarians. Biomimicry was taken into account on different levels including the use of a microstructure and the shape of the microstructure, as well as the featuring of a roughened surface. Finite element modelling was employed to adapt and adjust the microstructure to a given load resulting in a lightweight design while appealing to an aesthetic sense (Fig. 4).

In the medical domain, namely in the design of permanent or temporary medical implants, researchers have discovered that the implant's performance is optimum when the implant presents a geometric biomimetic design where minimum material is used. Not only have researchers managed to design lightweight structures, but

Fig. 3 New acetabular component with graded cellular section [60]

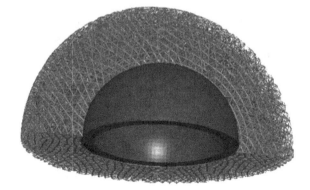

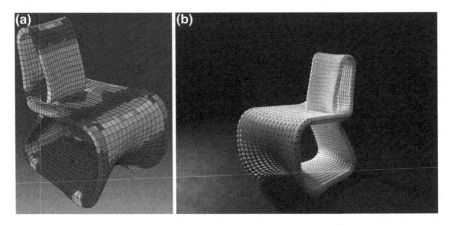

Fig. 4 **a** Numerical simulation model [64] and **b** photograph of the produced chair (http://us. archello.com/en/product/cellular-loop-chair-based-biomimetics#)

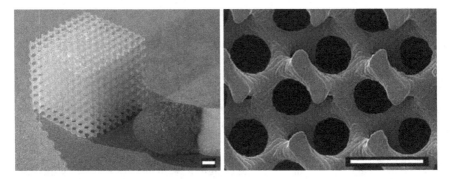

Fig. 5 Scaffolds with a gyroid architecture built by stereolithography. *Scale bars* represent 500 μm [41]

they have also managed to mimic the mechanical behaviour of the natural tissues, increasing the structural compatibility. These developments are possible only due to the parallel development of bioengineering and additive manufacturing.

Melchels et al. [41] used a resin based on poly (D,L-lactide) macromonomers and nonreactive diluent to produce porous scaffolds with gyroid architecture (Fig. 5). It was also possible to observe that preosteoblasts readily adhered and proliferated homogeneously on these scaffolds.

Diego et al. [18] characterised a specific family of scaffolds based on a face-cubic centred (FCC) arrangement of empty pores, leading to analytical formulae of porosity and specific surface (Fig. 6). The effective behaviour of these scaffolds was evaluated in terms of their mechanical properties and permeability, through the asymptotic homogenisation theory applied to a representative volume element identified with the unit-cell FCC [52].

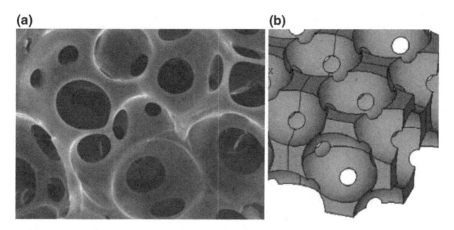

Fig. 6 **a** and **b** Comparison of a scanning electron microscope image of bone tissue and a scaffold based on the FCC design. Adapted from Diego et al. [18] and Sanz-Herrera et al. [52]

Whatley et al. [61] developed a biomimetic elastic intervertebral disk scaffold using a custom-designed 3D additive manufacturing device. To mimic the elastic nature of the native intervertebral disk tissue, a biodegradable polyurethane material was used for the elastic scaffold. Also, due to the proposed additive manufacturing system, a concentric lamellar structure mimicking the natural shape of the intervertebral disk was printed (Fig. 7).

Murr et al. [44] fabricated an exaggerated tibial–knee stem with various density compatibilities representing the high-end trabecular regime (approx. 0.8 g/cm^3) and the low-end cortical bone regime (approx. 1.5 g/cm^3; [15, 17]). Figure 8 shows the fabricated prototypes with varying outer mesh densities (left to right) of 0.86, 1.22, and 1.59 g/cm^3. An enlarged inset for the 0.86 g/cm^3 mesh is also shown.

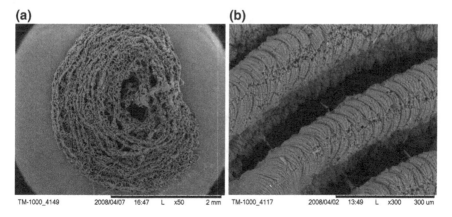

Fig. 7 **a** 3D scaffold structure; **b** multiple layers of polyurethane stacked in a 3D structure [61]

Fig. 8 Knee implant (tibial stem) prototype development by electron beam melting. Increasingly dense mesh array stems from the right to left [44]

In addition to producing lightweight products, biomimetic design may also be used for the production of tools for manufacturing purposes. Au and Yu [1, 2] proposed a porous cooling passageway to achieve conformal cooling. The porous structure used hollow hexagonal prisms, based on the structure of the honeycomb, and its assembly for developing the conformal cooling passageway (Fig. 9). The designed porous configuration cooling passageway not only provided an optimised cooling for the plastic injected part, but also increased the rate of heat transfer, while still providing the necessary mechanical strength to support the injection process. Short cycle times and good plastic product quality can thus be ensured.

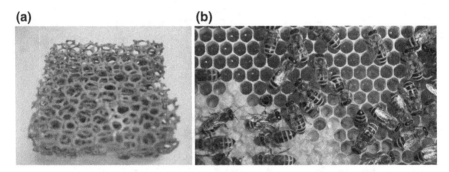

Fig. 9 a Scaffold elements with porous structure arrangement; **b** structure of a bee honeycomb [1, 2]

Strano et al. [57] proposed a new approach to the design of support structures that optimise the part built orientation and the support cellular structure. This approach applied a new optimisation algorithm to use periodic implicit functions for the design and generation of the cellular support structures including graded supports. Different cellular structures could be easily defined and optimised in order to have graded structures providing more robust support where the object's weight concentrated and less support elsewhere. Evaluation of a support optimisation scheme for a complex shape geometry revealed that the new approach presented can achieve significant material savings, thus increasing the sustainability and efficiency of metallic additive manufactured parts.

3 Topology Optimisation

The concept of designed cellular materials is motivated by the desire to allocate material only where it is needed for a specific application. From a mechanical point of view, the key issue offered by cellular materials is the high strength accompanied by low mass [26]. When a cellular design is not capable of being implemented in a specific product for a high strength/low mass ratio, optimisation methods may be used in order to overcome these limitations. Three optimisation methods have been explored for additive manufacturing applications, namely size or parametric, shape, and topological [20, 26, 39, 58]. In size optimisation, the dimensional values (such as thickness, diameters, etc.) are modified, whereas in shape optimisation, the boundaries of the shapes of the part are modified in order to change the domain of the structure aiming to achieve the optimal layout. In the topological optimisation method, the distribution of material is explored with the objective of obtaining a product with a high strength/low mass ratio, similar to the behaviour of natural cellular structures.

According to Zegard and Paulino [63], optimal structures can be designed through three topological optimisation techniques, namely ground structure optimisation, either 2D or 3D, and density-based topology optimisation.

The first optimisation technique, the ground structure method [19], numerically approximates the optimal trusslike structure [30, 43] using a reduced finite number of truss members. This technique eliminates unnecessary trusses from a highly interconnected truss-based structure while keeping the nodal locations fixed. In other words, the method avoids the optimisation of sizing and geometry by instead doing a sizing-only optimisation for a highly redundant truss. Hegemier and Prager [29] showed that a truss of a single load with maximum stiffness is fully stressed, meaning that volume minimisation with equal stress limits in tension and compression is equivalent to a compliance minimisation with a prescribed volume [6].

The third optimisation technique, density-based topology optimisation, is a method that tries to answer, 'Which is the best distribution of material for a given set of solicitations and constraints for a given design region?' The method performs by discretising the design domain and optimising density variables associated with each element within the finite element meshed discretisation domain. With this methodology, it can be determined if that material point should be full of material or void. Thus, the density of the element is used to manipulate the element stiffness and mass matrices which subsequently drives the objective function value. An example of an objective function is minimising compliance, in other words, achieving maximum global stiffness when subjected to a constraint on the amount of material that can be used [6, 39]. One of the most common density-based optimisation methods is the solid isotropic material with penalisation (SIMP) method. It is a gradient-based iterative method that performs in each iteration design updates and evaluations until a structural equilibrium is implicitly satisfied [6, 16].

Within the additive manufacturing domain, several authors have used and implemented topological optimisation methods: (1) for the design of the final part, (2) the design of the internal patterns of the final part, (3) as well as the support structures necessary during the building process of the final part. In either case, the objective is to reduce the amount of material needed while maintaining an optimal mechanical performance. An example is provided for each case.

From the three, it is in product design where topological optimisation has been more implemented. In the medical field, by applying topological optimisation techniques in the design process, not only do the biomedical engineers design a lightweight implant with a lower amount of raw material needed for the desired implant, but also the design is mechanically more robust. Zegard and Paulino [63] presented a case study where topology optimisation was used for the design of a maxilo-facial implant. Another example is in the aerospace industry with the design of the Airbus A320 nacelle hinge bracket presented by Gardan and Schneider [20] (Fig. 10).

Zegard and Paulino [63] explored the combination of both optimisation methods, the ground structure method and the density-based method using SIMP as illustrated in Fig. 12. The authors state that each method has strengths and weaknesses, but when combined provide valuable information of the optimal structural mechanism, while providing a wide range of design options (Fig. 11).

Gardan and Schneider [20] implemented topological optimisation methods to design the internal patterns of additive manufacturing products, allowing the conservation of the outer skin and the optimisation of the inner part of the product, as illustrated in Fig. 13. To validate their methodology, the design of a prosthetic implant used in surgical hip replacement procedures was studied. It is possible to observe the inner part of the hip implant in the design phase and after production. The weight gain is, lower material usage is achieved, and the methodology was tested in more than 10 parts produced from different additive manufacturing

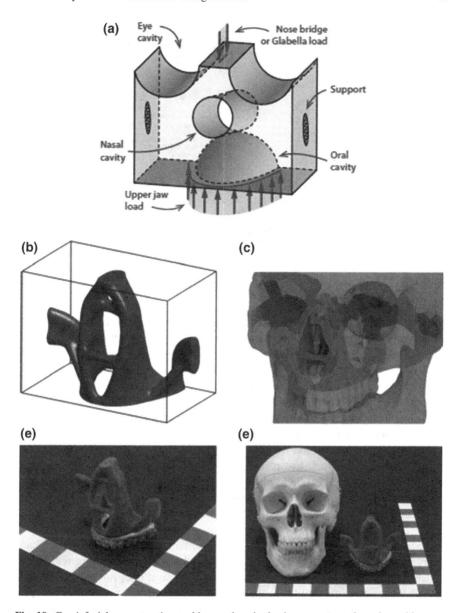

Fig. 10 Craniofacial reconstruction problem: **a** domain, loads, supports, and passive-void zones; **b** topology optimised result; **c** rendering of the resulting optimised topology embedded in a human skull; **d** manufactured model using a fused deposition modelling system with a cast of the main author's teeth for reference [scale in inches]; **e** manufactured model using a fused deposition modelling system with a human skull replica as reference [scale in inches] [63]

Fig. 11 Airbus A320 nacelle hinge bracket (*back*) and the optimised design produced by ALM (*front*). Courtesy of EADS and ALTAIR [20, 63]

processes, namely fused deposition modelling, selective laser sintering, and stereolithography. Gardan and Schneider [20] also used their methodology for the design of the support structures during production (Fig. 14).

Leary et al. [37] used topological optimisation methods to ensure manufacturability of additive manufacturing products without requiring additional support material. According to the authors, by assessing the manufacturing time and component mass associated with feasible orientations of the proposed geometry, an optimal orientation can be identified. Figure 15 illustrates the implemented design methodology on the extrusion-based produced parts.

The combination of additive manufacturing technologies and topological optimisation schemes may also be used to design a family of customised, lightweigh, and complex products [46, 49]. A product family design involves the concurrent design of multiple products based on a common product platform to satisfy a variety of markets [54]. The convenient balancing between the commonality and the performance of each individual product in the family is an issue in the product family paradigm. Lei et al. [46] considered that additive manufacturing will bring fundamental changes in the design of family products as it does not require tooling and has great design flexibility. In their work, Lei and Moon presented a product family design methodology to create sustainable products, consisting of a topology optimisation followed by a finite element analysis. The boundary conditions for the topology optimisation step translate the individual product requirements into distinct object structures. Therefore, each of these objects constitutes an individual component which is unique for each product and is optimal in terms of material use.

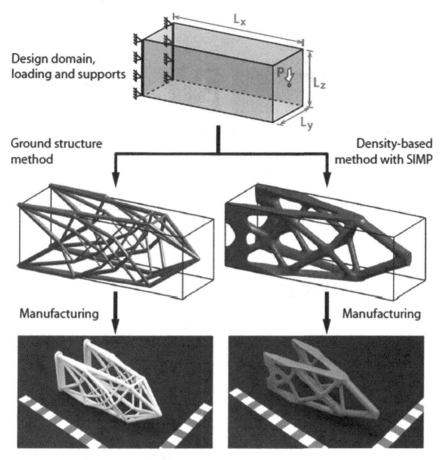

Fig. 12 Two distinct (but related) optimal solutions to be obtained and manufactured: a ground structure (trusslike) solution and a density-based (continuum) solution [63]

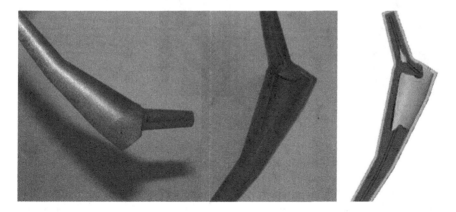

Fig. 13 View of manufactured hip right out of the oven [20]

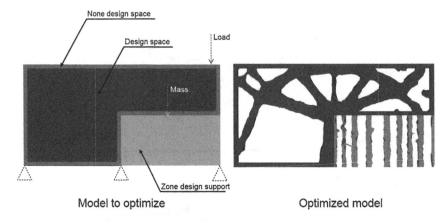

Fig. 14 Simple example of part and support optimisation [20]

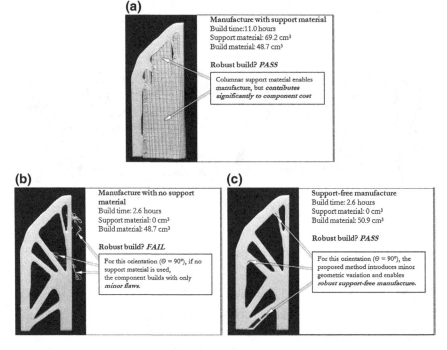

Fig. 15 Production of a part with **a** support material, **b** no support material, and **c** optimised without supports [37]

Through the use of additive manufacturing, these optimized components consume less material and less energy, therefore they are more cost effective and more sustainable than a component which is common to all products in the family. This approach can be compared to what happens in nature, where a unique component can be found to obey different functions in the same organism depending on whether it is associated with other components. For example, a protein, such as type I collagen, presents different morphologies in different tissues to perform different functions depending on its association with hydroxyapatite crystals: it gives rigid (high Young modulus) and shock-resistant tissues in bone, it behaves as an elastomer with low rigidity and high deformation to rupture in tendons, or shows optical properties such as transparency in the cornea [51].

4 Conclusions

Since life itself began on Earth, nature has evolved in such a form that it has reached optimised biological forms, designs, functions, and concepts. As a result of this evolution, nature has solved engineering problems such as self-healing abilities, environmental exposure tolerance and resistance, surface performance (e.g., hydrophobicity, self-cleaning, high and reversible adhesion), self-assembly, and energy conversion and conservation, all in a sustainable manner.

Based on these design principles, among the existing manufacturing technologies, additive manufacturing is the most promising, innovating way of producing components and possessing good sustainable characteristics, such as utilising only the amount of material needed for the building of the final part, by reducing the material mass and energy consumption, eliminating scrap and the need for tooling, and also the use of environmentally damaging process enablers.

One of the possible solutions is to become biologically inspired in order to produce optimum parts with all their required properties and functionalities while reducing the amount of raw material needed for production. In order to demonstrate this biomimetic design inspiration, several case examples have been presented, either by considering cellular structures or by considering topological optimisation schemes. From the given examples, it is possible to observe that biomimicry is a fundamental tool to design parts and tools with optimum performance while reducing the amount of raw material.

Acknowledgments This work has been supported by the Fundação para a Ciência e a Tecnologia (FCT) under project grant UID/MULTI/00308/2013.

References

1. Au KM, Yu KM (2005) Porous cooling passageway for rapid plastic injection moulding. In: Bartolo P et al (eds) Virtual modelling and rapid manufacturing. Taylor & Francis, London
2. Au KM, Yu KM (2007) A scaffolding architecture for conformal cooling design in rapid plastic injection moulding. Int J Adv Manuf Technol 34(5–6):496–515
3. Baumers M (2012) Economic aspects of additive manufacturing: benefits, costs and energy consumption. Doctoral thesis, Loughborough University, Leices-tershire, United Kingdom

4. Baumers M, Tuck C, Wildman R, Ashcroft I, Hague R (2011) Energy inputs to additive manufacturing: does capacity utilization matter?. In: Conference paper: solid freeform fabrication symposium 2011, Austin, TX, USA
5. Beaman JJ, Barlow JW, Bourell DL, Crawford RH, Marcus HL, McAlea KP (1997) Solid freeform fabrication: a new direction in manufacturing. Kluwer Academic Press, Boston
6. Bendsøe MP, Sigmund O (2004) Topology optimization: theory, methods and applications. Springer, Berlin
7. Benyus J (2002) Biomimicry: innovation inspired by Nature. HarperCollins, New York
8. Berry M (2004) The importance of sustainable development. Columbia Spectator, Canada
9. Bhushan B (2009) Biomimetics: lessons from Nature—an overview. Phil Trans R Soc A 367:1445–1486
10. Bourell DL, Leu MC, Rosen DW (2009) Roadmap for additive manufacturing: identifying the future of freeform processing. The University of Texas, Austin
11. Bourhis FL, Kerbrat O, Hascoet J, Mognol P (2013) Sustainable manufacturing: evaluation and modeling of environmental impacts in additive manufacturing. Int J Adv Manuf Technol 69(9):1927–1939
12. Bourhis FL, Kerbrat O, Dembinski L, Hascoet JY, Mognol P (2014) Predictive model for environmental assessment in additive manufacturing process. Procedia CIRP 15:26–31
13. Campbell RI, Hague RJ, Sener B, Wormald PW (2003) The potential for the bespoke industrial designer. Des J 6(3):24
14. Campbell T, Williams C, Ivanova O, Garrett B (2011) Could 3D printing change the world? Technologies, and implications of additive manufacturing. Atlantic Council, Washington, DC, USA
15. Cezayirlioglu H, Bahnivk E, Davy DT, Heiple KG (1985) An isotropic behavior of bone under combined axial force and torque. J Biomech 18:61–69. doi:10.1016/0021-9290(85)90045-4
16. Christensen P, Klarbring A (2009) An introduction to structural optimization, 1st edn. Springer, Berlin
17. Currey JD (2002) Bones: structure and mechanics. Princeton University Press, Princeton
18. Diego RB, Estellés JM, Sanz JA, García-Aznar JM, Sánchez MS (2007) Polymer scaffolds with interconnected spherical pores and controlled architecture for tissue engineering: Fabrication, mechanical properties and finite element modelling. J Biomed Mater Res B 81B:448–455
19. Dorn WS, Gomory RE, Greenberg HJ (1964) Automatic design of optimal structures. J Mecanique 3(1):25–52
20. Gardan N, Schneider A (2014) Topological optimization of internal patterns and support in additive manufacturing. J Manuf Syst. doi:10.1016/j.jmsy.2014.07.003
21. Gebler M, Uiterkamp AJMS, Visser C (2014) A global sustainability perspective on 3D printing technologies. Energy Policy 74:158–167
22. Gibson I, Rosen DW, Stucker B (2010) Additive manufacturing technologies: rapid prototyping to direct digital manufacturing. Springer, New York
23. Gibson LJ (2005) Biomechanics of cellular solids. J Biomech 38:377–399
24. Gibson LJ, Ashby MF (1997) Cellular solids: structure and properties, 2nd edn. Cambridge University Press, Cambridge
25. Gibson LJ, Ashby MF, Karam GN, Wegst U, Shercliff HR (1995) The mechanical properties of natural materials II: Microstructures for mechanical efficiency. Proc R Soc Lond A450:141–162
26. Gibson I, Rosen D, Stucker B (2015) Additive manufacturing technologies—3D printing, rapid prototyping and direct digital manufacturing, 2nd edn. Springer, New York
27. Hague R (2005) Unlocking the design potential of rapid manufacturing. In: Hopkinson N et al (eds) Rapid manufacturing: an industrial revolution for the digital age. Wiley, New York
28. Hao L, Raymont D, Yan C, Hussein A, Young P (2011) Design and additive manufacturing of cellular lattice structures. In: Bartolo P et al (eds) Innovative developments in virtual and physical prototyping. Taylor & Francis, London
29. Hegemier G, Prager W (1969) On Michell trusses. Int J Mech Sci 11(2):209–215

30. Hemp WS (1973) Optimum structures, 1st edn. Oxford University Press, Oxford
31. Hopkinson N, Gao Y, McAfee DJ (2006) Design for environment analyses applied to rapid manufacturing. Proc Inst Mech Eng D J Automob Eng 220(D10):1363–1372. doi:10.1243/09544070jauto309
32. Howarth G, Hadfield M (2006) A sustainable product design model. Mater Des 27:1128–1133
33. Huang SH, Liu P, Mokasdar A, Hou L (2013) Additive manufacturing and its societal impact: a literature review. Int J Adv Manuf Technol 67:1191–1203
34. Kowalczyk P (2003) Elastic properties of cancellous bone derived from finite element models of parameterized microstructure cells. J Biomech 36:961–972
35. Kreiger M, Pearce JM (2013) Environmental life cycle analysis of distributed three-dimensional printing and conventional manufacturing of polymer products. ACS Sustain Chem Eng 1(12):1511–1519
36. Krishnakumar V (2012) Biomimetic architecture, seminar 2011–2012. School of Planning and Architecture, National Institute of Technology Calicut, India
37. Leary M, Torti LMF, Mazur M, Brandt M (2014) Optimal topology for additive manufacture: a method for enabling additive manufacture of support-free optimal structures. Mater Des 63:678–690
38. Luo Y, Ji Z, Leu MC, Caudill R (1999) Environmental performance analysis of solid freeform fabrication processes. In: Proceedings of the 1999 IEEE international symposium on electronics and the environment (ISEE-1999), IEEE
39. Marchesi TR, Lahuerta RD, Silva ECN, Tsuzuki MSG, Martins TC, Barari A, Wood I (2015) Topologically optimized diesel engine support manufactured with additive manufacturing. IFAC-Pap 48(3):2333–2338
40. Masters M, Mathy M (2002) Direct manufacturing of custom made hearing instruments, an implementation of digital mechanical processing. Paper presented at the SME rapid prototyping conference and exhibition, Cincinnati, OH, USA
41. Melchels FP, Feijen J, Grijpma DW (2009) A poly (D, L-lactide) resin for the preparation of tissue engineering scaffolds by stereolithography. Biomaterials 30:3801–3809
42. Meyers MA, Chen P-Y, Lin AY-M, Seki Y (2008) Biological materials: Structure and mechanical properties. Prog Mater Sci 53:1–206
43. Michell AGM (2010) The limits of economy of material in framestructures. Philos Mag Ser 6 8(47):589–597
44. Murr LE, Medina SMGF, Lopez H, Martinez E, Machado BI, Hernandez DH, Martinez L, Lopez MI, Wicker RB, Bracke J (2010) Next-generation biomedical implants using additive manufacturing of complex, cellular and functional mesh arrays. Phil Trans R Soc A 368:1999–2032. doi:10.1098/rsta.2010.0010
45. Nguyen J, Park SI, Rosen DW (2011) Cellular structure design for lightweight components. In: Bartolo P et al (eds) Innovative developments in virtual and physical prototyping. Taylor & Francis, London
46. Lei N, Moon SK, Bi G (2013) Additive manufacturing and topology optimization to support product family design. In: Bartolo P et al (eds) High value manufacturing. Taylor & Francis, London
47. Petrovic V, Gonzales JVH, Ferrado OJ, Gordillo JD, Puchades JRB, Ginan LP (2011) Additive layered manufacturing: sectors of industrial application shown through case studies. Int J Prod Res 49(4):1071–1079
48. Prakash WN, Sridhar VG, Annamalai K (2014) New product development by DFMA and rapid prototyping. ARPN J Eng Appl Sci 9(3):274–279
49. Rosen DW (2007) Computer-aided design for additive manufacturing of cellular structures. Comput Aided Des Appl 4(5):585–594
50. Rossin KJ (2010) Biomimicry: Nature's design process versus the designer's process. WIT Trans Ecol Envir 138
51. Sanchez C, Arribart H, Guille MMG (2005) Biomimetism and bioinspiration as tools for the design of innovative materials and systems. Nat Mater 4(4):277–288

52. Sanz-Herrera JA, Garcia-Aznar JM, Doblaré M (2009) On scaffold designing for bone regeneration: a computational multiscale approach. Acta Biomater 5:219–229
53. Serres N, Tidu D, Sankare S, Hlawka F (2011) Environmental comparison of MESO-CLAD® process and conventional machining implementing life cycle assessment. J Clean Prod 19(9–10):1117–1124
54. Simpson TW, Seepersad CC, Mistree F (2001) Balancing commonality and performance within the concurrent design of multiple products in a product family. Concurrent Eng 9(3):177–190
55. Sreenivasan R, Goel A, Bourell DL (2010) Sustainability issues in laser-based additive manufacturing. LANE 2010. Phys Procedia 5:81–90
56. Stampfl J, Cano Vives R, Seidler S, Liska R, Schwager F, Gruber H et al (2003) Rapid prototyping—a route for the fabrication of biomimetic cellular materials. In: Proceedings of the 1st international conference on advanced research in virtual and rapid prototyping, pp 593–599
57. Strano G, Hao L, Everson RM, Evans KE (2013) A new approach to the design and optimisation of support structures in additive manufacturing. Int J Adv Manuf Technol 66:1247–1254. doi:10.1007/s00170-012-4403-x
58. Thakur A, Banerjee AG, Gupta SK (2009) A survey of CAD model simplification techniques for physics-based simulation applications. Comput Aided Des 41:65–80
59. Wang HV (2005) A unit cell approach for lightweight structure and compliant mechanism. Doctoral dissertation, Georgia Institute Of Technology
60. Wang HV, Johnston SR, Rosen DW (2006) Design of a graded cellular structure for an acetabular hip replacement component. In: 17th solid freeform fabrication symposium, vol 238, p 111
61. Whatley BR, Kuo J, Shuai C, Damon BJ, Wen X (2011) Fabrication of a biomimetic elastic intervertebral disk scaffold using additive manufacturing. Biofabrication 3(1):015004
62. Yang S, Zhao YF (2015) Additive manufacturing-enabled design theory and methodology: a critical review. Int J Adv Manuf Technol 80:327–342
63. Zegard T, Paulino GH (2015) Bridging topology optimization and additive manufacturing. Struct Multidisc Optim. doi:10.1007/s00158-015-1274-4
64. Ziegler T, Jaeger R (2012) A light weight design approach for aesthetic consumer goods using biomimetic structures In: Demmer A (Hrsg.) (eds) Proceedings of Fraunhofer direct digital manufacturing conference DDMC 2012. Fraunhofer IPT, Aachen

Sustainable Frugal Design Using 3D Printing

Ian Gibson and Abhijeet Shukla

Abstract This chapter describes and discusses achieving sustainable solutions through the fusion of 3D printing with frugal approaches in design and engineering. Sustainability has various definitions, however. According to the widely accepted Bruntland Report (World Commission on Environment and Development in Our common future. Oxford University Press, Oxford, p. 27, 1987) [1] for the World Commission on Environment and Development (1992), sustainability can be defined as 'Development that meets the needs of the present without compromising the ability of future generations to meet their own needs'. We can use this definition when we are dealing with product design to ensure the design functions correctly without causing a severe impact on available resources. For the chapter we consider sustainability as obtaining viable and tangible solutions for a given design problem which are environmentally friendly and economically advantageous for end users and society. 3D printing (3DP) is often regarded as a 'disruptive' technology that has forced many to rethink how we design and make things and how to turn this into new business models. It is also often considered as a 'liberating' technology that is easier to use, enabling us to transfer ideas generated in a digital format into physical forms with minimal fuss and cost. This chapter explores the product design and development process in relation to frugal and sustainable concepts and how 3D printing and related technologies can influence them.

Keywords Sustainable frugal design · 3D printing · Energy · Material utilization · Tool design

I. Gibson (✉)
School of Engineering, Deakin University, Waurn Ponds, Australia
e-mail: ian.gibson@deakin.edu.au

A. Shukla
Indian Institute of Technology, Kanpur, India

© Springer Science+Business Media Singapore 2016   85
S.S. Muthu and M.M. Savalani (eds.), *Handbook of Sustainability
in Additive Manufacturing*, Environmental Footprints and Eco-design
of Products and Processes, DOI 10.1007/978-981-10-0606-7_4

1 Introduction

We start by analysing design in terms of the resources used during the process and in relation to the designs generated. Resources come in a variety of forms, but probably the most important ones used in design are manpower, energy, materials, and technology.

Manpower can involve the designers themselves, but also support staff, particularly in relation to the building of prototypes. It is generally measured in terms of time committed to the design process and there have been many attempts to optimise, automate, and generally speed up this process.

Excessive use of *Energy* is not usually an issue during the design process as we are often dealing with small numbers of products being made during this time. However, energy does become a focus for designers when they are considering the product's consumption during use as well as the energy required during manufacture.

Materials can be considered in a similar way to energy. The product design and prototyping process can be quite wasteful in terms of materials use because it is often necessary to build something in order to evaluate functionality. This would then allow the product to develop and therefore the material used would often be discarded. However, as mentioned above, we are dealing with small numbers and greater efficiencies can be had in downstream manufacturing stages.

The use of computer *technology* has revolutionised the design process. Many designs can be evaluated to a much greater extent as software models in some form of simulation. Although technology is primarily aimed at increasing efficiency, it can also reduce consumption of time, energy, and materials. Furthermore, there are downstream impacts where manufacturing technology can make better use of the digital information generated during this design process.

When considering frugal approaches to design we must understand it from both the product and process perspectives. The above-mentioned process issues relate mostly to the design process. A product perspective can be most easily understood through case examples. With increasing population and depleting natural resources, it is important to educate our designers on the important role they play in ensuring a more sustainable future. Designers must be frugal in the way they design and the products they design. Effective use of technology is part of this education.

2 Categorisation of Frugal Design

When considering frugal design, it is worth noting that there may be different approaches connected to the type of design. For example, we may like to look at frugal design of:

(a) Medical products
(b) Social innovations

(c) Business innovations
(d) Approaches in education and learning
(e) Aerospace technology
(f) Products and processes with environmental benefit
(g) Consumer products.

Medical products must have value to treat health issues or to maintain a good standard of health and this must be considered a priority during the design process. Poorly conceived or executed designs could potentially cause more harm than good. However, we can consider frugality at many levels in the development of products. The most obvious consideration would be to look at products that have been generated using a traditional approach and then redesigned to suit a poorer social group or a larger market to suit social and business requirements. Conventional approaches generally require large investment and development programmes that ensure safety and correct certification. Large companies don't normally consider sustainability during this process, at least not for new product innovations. However, it is worthwhile considering, even for complex aerospace applications where frugal approaches can result in significant operational savings, such as fuel costs. There is therefore significant value to incorporate frugality and sustainable thinking into grass-roots education so that all consumers may be able to benefit.

As part of an experiment in frugal design thinking, we can consider how a technology such as 3D printing can have influence related to the above categories. We revisit these topics once we explain a little more clearly what 3DP actually is to see how we can use it to make a more sustainable world.

3 3D Printing

3D printing (3DP) is the popular name for what is technically termed additive manufacturing and what has also been referred to as freeform fabrication, rapid prototyping, and layer-based manufacturing [2]. 3DP is generally compared with traditional carving and machining-based fabrication methods where material is removed (in a subtractive manner) to reveal the form of the final part. Instead, 3DP requires the addition of material, usually in a layerwise manner, to build up the final part. If the layers are sufficiently thin, then they can be considered as finite-thickness 2D cross-sections. This simplification of a complex 3D problem into a series of simpler 2D problems enables the process to be driven easily from computer-generated models. With the proliferation of 3D models and the ease in which they can be made, 3DP has proven to be a popular method of conversion from virtual to physical form. Further to this, 3DP printing machines have come dramatically down in price, making them affordable to individual users for their own recreational purposes.

The process for 3DP starts with a 3D digital geometric model. This model is usually processed within computer-aided design (CAD) software, but it is increasingly common for the source of this 3D model to come from medical scanning data or from 3D scanning technology. Such datasets allow the models to be customised to suit an individual or a specific location.

The real benefit of 3DP is that the process of achieving a real physical part is short, straightforward, and independent of the part geometry that is to be created. Figure 1 shows a typical process chain. Once the geometry has been specified by the user, the normal practice is to convert the file data into a generic form called an STL file. This file is a 3D surface model that can be easily used by the 3DP process to generate slice data. At this point, the machine may require some material or process-specific information such as slice thickness, build temperature, and the like. Default parameters can also be used here. Some processes require the generation of a supporting structure to be created to hold the part in place during the build. These supports would need to be removed once this build is complete.

The fact that parts are built using layers of material implies that the final part is an approximation of the intended geometry that is defined by the thickness of the layers. For certain applications, this approximation may be acceptable without further action. However, many applications would require the resulting parts to be post processed in order to achieve a smoother surface. Surface coatings may also be required to provide a desired appearance that cannot be achieved inside the 3DP machines. It is, however, quite easy to see that moving from art (computer model)

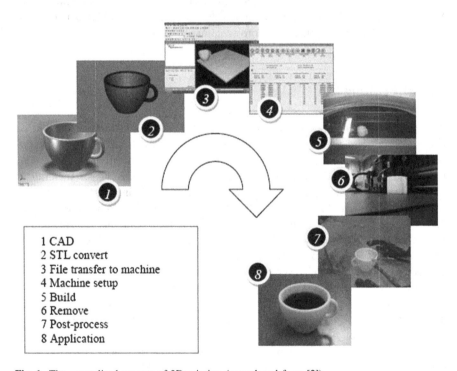

1 CAD
2 STL convert
3 File transfer to machine
4 Machine setup
5 Build
6 Remove
7 Post-process
8 Application

Fig. 1 The generalised process of 3D printing (reproduced from [2])

to part (physical model) is a systematic, largely automated, and straightforward process.

Applications can often be categorised according to the generic requirements from some of the early adopting industries:

Automotive industries have used 3DP primarily in the product development process. Here, designers are looking to create prototypes that will allow them to learn about how the final parts will behave in order to achieve a rapid time to market.

Aerospace industries use the technology to achieve final parts that have complex geometries. As previously mentioned, CAD modelling can produce surface geometry that can be difficult to fabricate using conventional means. For example, 3DP can create parts with features internal to the structure.

Medical industries are looking to create devices that fit to the unique form of an individual patient in a quick and affordable manner. By importing medical scan data, it is possible for 3DP to fabricate such geometries.

It is however becoming apparent that there is a fourth class of user. This generic class can be termed *innovators* who are using 3DP not only to develop novel geometric structures but also new business models. Media also refer to this group as 'makers' or 'disruptors' and it is clear that although 3DP is a key component in much of what they do, they also take advantage of other forms of simple-to-use, versatile technologies and combine them to create new and novel solutions. This integration of technologies is happening because they have achieved a stage of maturity where they are easy to use, fairly reliable, and affordable. These technologies range from computer- and mobile-based systems, through laser cutting/etching, to graphics and 3D scanning technology. We all know that mobile smartphones have onboard technology that can detect position, orientation, light levels, temperature, and other environmental conditions. Different application developers can use the same device to perform a variety of functions. This 'technology convergence' can also be used by designers to create 'frugal' solutions by adding a small amount of high-tech to low-tech systems to provide innovative solutions to everyday problems.

4 Frugal Design and Engineering

Frugality in design and engineering can be understood as the approach towards a desired target by using minimal available resources and eliminating inefficient factors without compromising the key features (as shown in Fig. 2). The concept of frugality may be traced from ancient India where in Sanskrit, the term *Yukti* is used for the innovative idea of providing a swift, unconventional, and alternative approach of solving a problem which could serve the purpose. The equivalent English term is known as 'frugality'. The concept of frugality is famously known and discussed as frugal innovation, Gandhian engineering, and Asian innovation. It is important to note the Asian roots to these approaches as we typically understand

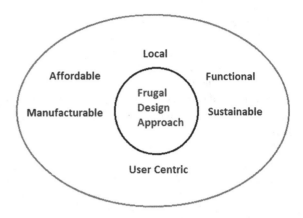

Fig. 2 Key features of frugal design approach

Asia to be a blend of high and low technology, illustrated by large wealth gaps and a significant population of very poor people combined with a wealthy infrastructure.

What makes the frugal approach more popular is the phenomenon of design to extreme affordability. Frugal design and engineering provides improvised disingenuous solutions combined with innovative 'smart' thinking. It can be understood as a philosophy of acting in a selfless, compassionate, but unconventional manner to create a better world for all, not just for a few. Simplicity of the frugal solution is one of the most sought-after criteria for solution selection; however, it is not necessary for any frugal approach to be simple.

Identifying the exact need is one of the important aspects of a frugal design approach. In the frugal approach a necessary/desirable output must be differentiated in smaller parts of achievable targets in a procedural format. Exact definition of need helps the designer to ideate in an efficient way and accurate achievable target differentiation helps in selecting the right approach. Frugal approach in design and engineering can be understood and practiced in the following areas.

- **Frugal Approach for Material Utilisation**

  This approach enables designers to use variations in material quantity and quality as per availability and requirement. The identified requirements shall be matched with available resources to perform. Let us discuss the case of 'adaptable baby carrier design (ABCD)' for developing countries. This example comes from a project that was part of a Design Programme held at IIT Kanpur as a course project for 'Special Studies in Design'. The objective of the project was to design and fabricate a low-cost baby carrier for developing countries which can be adapted by users much more easily than conventionally available products on the market. User survey and market research were performed by the team to validate and identify any problems other than those specified in the design brief. Cost and adaptability of the product are interrelated and linked by the material requirements for the product. The materials used in the existing products were costly and designs were intuitively not adaptable to a wide range

Fig. 3 Design of frugal baby carrier

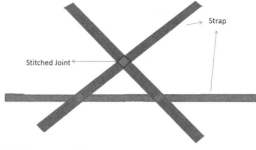

Material- Denim/Dungaree cloth
Material is planned to recycle from used Jeans which are discarded by users

of users (parents/baby holder). However, the ergonomics and baby postures provided by the existing products were well researched and validated by the companies providing the solution. To achieve adaptability in the new design solution and also cost reduction through utilisation of material, a concept was developed fitting the context of frugal approach in material utilisation.

In Concept 1 shown in Fig. 3 denim fabric strips from used jeans and denim suits were planned to upcycle and stitched together to form a baby carrier which can be tied at the back or in front of the user to carry the baby. Figure 4 shows a volunteer using the paper-made prototype to carry a dummy baby at his back. The frugal approach in selection of material and utilisation for the baby carrier lowered the cost of the product by up to 80 % compared to the market price of conventional baby carrier products. The frugal approach in this case has provided a sustainable solution for used and discarded fabric to re-enter the market and extend the utilisation period against a favourable cost of production.

- **Frugal Approach for Tool Design**

  Available materials can often be reformed and utilised as tools for specific purposes. Shape, size, material properties, and strength of material can be

Fig. 4 Mockup of frugal baby carrier [3]

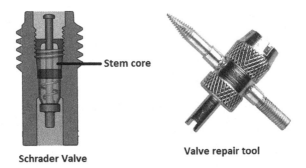

Fig. 5 Typical removal tool for Schrader valve repair

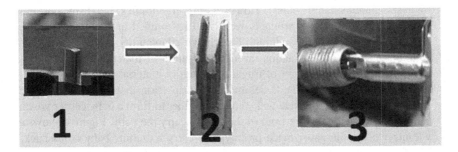

Fig. 6 Valve stem core opener

considered as parameters for identifying assets that may be utilised as such tools. There may also be some changes required for the fitting of selected objects to perform the tooling tasks; these changes can be permanent as well as temporary in nature depending on the task and physical properties of the material. Let us discuss the frugal approach for a tool design case in a design problem of cleaning a Schrader valve usually found in bike and car tyre tubes. To open the poppet assembly of the valve, a special tool is used as shown in Fig. 5.

But it is not necessary for every person, designer, engineer, and maker to have all the tools available every time in his or her workshop. Buying the tool can be an option but a tool can also be frugally designed by simply looking into available resources and finding something close to the requirement and then reshaping it to use as a tool to solve the purpose. As shown in Fig. 6 a simple metallic rod is selected to be reformed as a valve stem core opener. It requires the simple process of cutting and grinding to be transformed into the required tool.

- **Frugal Approach in Methods and Processes**

  In this approach a product, process, method, or system is identified and utilised to perform for desired process parameters for which the chosen system, product, process, and/or method is not designed. It can be achieved by merging and/or eliminating subsystems from other systems or the host. The process or method desired to be achieved should be economical.

Fig. 7 CNC machine to 3D printer [4]

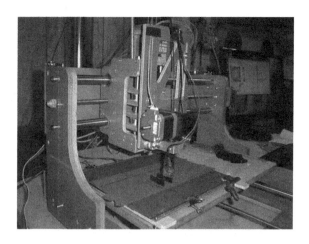

Conversion of three-axis CNC machines to a 3D printer is a good example of this approach. The CNC machine is a subtractive manufacturing unit and is totally different from the method and process of additive manufacturing; however, CNC shares a common physical three-axis mechanism for moving and positioning the tool from one point to another. As shown in Fig. 7, an FDM (fused deposition modelling) extruder is placed at the tool post of the CNC machine and STL to G-code converting software is used to get the modified code for 3D printing using CNC. The output from this setup should not be compared to the 3D printed part from a professional 3D printer. But this setup surely solves the purpose on a rough scale.

- **Frugal Approach in Application**

  Exploration of the product or process for in-genuine application is attempted in this approach. A computer-controlled two- or three-axis mechanism as shown in Fig. 8 can be made using CD drives of old computers. These mechanisms can be further used as plotter, CNC, and 3D printer. For old CD drives the application

Fig. 8 Three-axis mechanism made using old CD drive parts

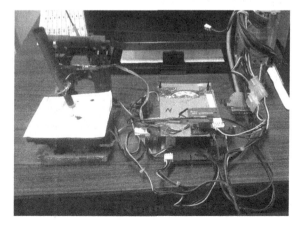

Fig. 9 Water bottle as funnel [5]

derived in this example is in-genuine all the way from design to application, as the designer of these CD drives did not design them to be used as a plotter or 3D printer.

- **Frugal Approach in Forms**
  Effective physical form according to requirements is generated in this approach. The form and shape of any other product can also be utilised in the approach to solve the purpose. Figure 9 shows a plastic water bottle can be utilised as a funnel by cutting it into the appropriate shape.
  A given or identified problem can be analysed for different parameters described above in the subheadings and then utilised for the concept generation for the solution of the problem. The whole process of frugal design can be understood by the flowchart representation in Fig. 10.

5 Frugal 3D Printing

As previously mentioned, 3D printing is often associated with innovative or disruptive business models. Such business models can include attention to frugal principles. The following are examples of how frugal approaches can be enabled by the use of 3D printing.

Component minimisation can be associated within frugal approaches. A product that has fewer components will generally be easier to assemble and maintain. The downside is that the components may be more difficult to manufacture. Drawing on the principle of 'complexity for free', 3DP may be able to overcome that obstacle. The example illustrated in Fig. 11 shows an air duct designed for an aircraft. Inasmuch as the duct has multiple channels and changes in direction, it is not possible to fabricate this easily as a single piece and the original solution required a number of components assembled together. Using a 3DP process that utilises an aerospace-grade material, it is possible to construct the entire duct, including mounting and connecting features, as a single piece. This is an excellent example to

Fig. 10 Flowchart for a frugal design process: *green* flow lines are positive responses, *red* flow lines are negative responses, and *blue* flow lines show loop

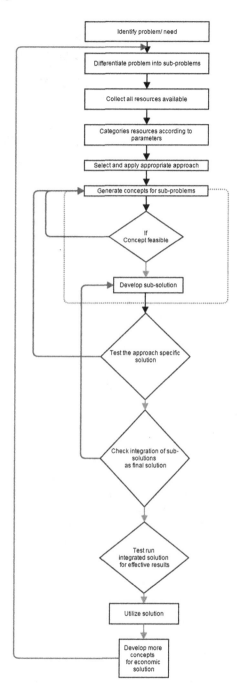

Fig. 11 Ducting redesign using 3DP [6]

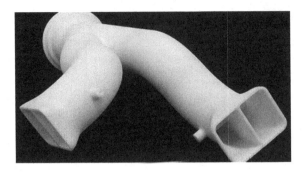

illustrate that 3DP can provide a frugal solution to complex, high-technology products as well as the more commonly-associated developing world issues. Note that a number of military organisations are using this approach to maintain their vehicles, thus proving that the quality and standard of such parts can meet stringent requirements in a range of existing applications.

Extending the previous concept further, it is possible to utilise 3DP to reduce inventory. This can be particularly relevant to spare-part storage but is also applicable to the manufacture of new parts. If it has been identified that 3DP can be used either to fabricate a new component or to replace an existing component, then it is possible to keep the inventory as a set of digital files. Immediately prior to use, we can initiate fabrication using 3DP to produce the required part. The data files will include part geometry, but may also contain information concerning the materials to be used, the required 3DP machine and build parameters, and other relevant supply-chain information.

Although aircraft and similar products have generic designs, there are often specific modifications that may make them unique. These modifications are referred to as *customisations* that adapt the base technology to suit a particular purpose or in response to the owner's or user's tastes. Manufacture of custom elements requires careful attention to detail and the site location of components. Design, planning, and tooling for these components can often be cost prohibitive. Noncustomised parts may fit the purpose to an extent but may often result in a suboptimal solution. Nowhere is this most evident than in the design and manufacture of medical devices. An excellent example where value is added through the use of 3DP to provide a higher performing result is in the design of in-the-ear hearing aids (Fig. 12). The casing of the hearing aid is taken from a digitised impression of a specific patient. This is then combined with more generic electronics and power to provide a more comfortable solution. This is now rapidly becoming the standard approach to hearing aid production, which also takes advantage of the complexity for free concept and the minimal inventory to provide a frugal solution.

To show that this approach of using 3DP to provide medical solutions can also extend into poor and deprived communities, the Robohand design provides an excellent illustration [8]. There can be numerous reasons why people can lose a hand. Sometimes, in isolated regions, it may be considered necessary or easier to

Fig. 12 Customised
in-the-ear hearing aids made
using 3DP [7]

amputate rather than save the hand. The Robohand device is one of a number of designs that have been developed to use low-cost 3DP to provide prosthetic solutions in this context (Fig. 13). The device can be modified easily to suit the size of the user and the extent of the injury and the components can then be scaled for building using 3DP. Furthermore, the design has a simple mechanism that uses cables and springs to open and close the prosthetic hand as the arm bends, thus providing a small element of dexterity. This design can therefore provide a significant additional value to the alternative, noncustomised, nonarticulated, strap-on systems. Many of these hands have been built and distributed through charitable and other sources. Examples have also been built for users in developed countries. It is worthwhile noting that many of these have no medical certification and have been built and (generally) very well accepted by the users and their communities because they are solutions that can be easily afforded.

Fig. 13 Robohand [8]

The Robohand and other low-cost solutions would not have come about if it had not been for a significant shift in the use and development in 3DP technology. In the mid to late 2000s, a number of key 3DP patents lapsed, particularly in relation to fused deposition modelling technology [ref Stratasys]. This allowed others to use this technology and create versions of FDM machines and sell them at much reduced prices. Many of these companies made use of an open-source version of the FDM technology called RepRap [9], an example of which can be seen in Fig. 14. By making this technology more affordable, the number of users has increased by a huge amount, vastly increasing the range of applications and allowing exploration into many new areas including food printing, bioprinting, and direct manufacture of low-cost, customised consumer goods.

Perhaps one of the most ambitious attempts to use 3DP to solve sustainability problems is in the automotive sector. An electric two-seater vehicle has been built, largely using a composite-material FDM-style printer. Shown originally as an experiment to be able to print a car in just two days, Local Motors has now launched the LM3D car shown in Fig. 15. Most of the chassis and body are made as a single piece using a carbon composite 3DP process. Although it still requires a lot of hand finishing and fitting out with motor, controls, and the like, this car can aim to replace the demand for low-cost short-range vehicles. Perhaps it is debatable whether it is a frugal solution, but one can see that the significant reduction in

Fig. 14 The RepRap Darwin machine [9]

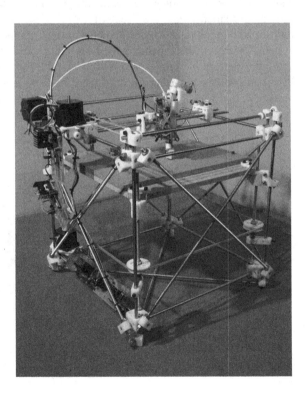

Fig. 15 LM3D car made using 3DP [10]

tooling, inventory, time, and manpower required to build it does indeed make it a potential contender. Regardless, it is an excellent illustration as to how far 3DP has come and how far it can go.

General issues that are worth mentioning at this stage include the ability for 3DP to make use of recycled material. Aside from the support material used to maintain the build geometry, 3DP only uses the material required to build the part. Comparing that with subtractive processes already makes it a material-efficient process. One must appreciate, however, that the material must generally be prepared according to a specific feedstock requirement and supplied in powder, filament, or cartridge form. The relatively low number of machines in circulation (when compared with CNC machines, e.g.) does mean that material is quite expensive. However, the low-cost FDM machines have driven material prices down and one can expect that to continue and eventually to cause this to happen for all 3DP technologies. It has been known for some users to produce their own feedstock, even using recycled materials. The problem with recycling polymers is that the mechanical properties are likely to degrade. Furthermore, it is not always clear what the original polymers were before recycling and one is very likely to be blending it with other materials with uncertain results. The common method to overcoming some of these problems is to use a filler material, such as chalk or rubber particles (taken from recycled car tyres) to create a more uniform result that may at least provide a reasonable compressive strength for nonload critical applications.

Another debatable point is energy utilisation. Inventory (and thus waste) is significantly reduced by using 3DP to create parts to order. If the 3DP machines are placed geographically close to the final market, then there would be significant reductions in shipping and the associated carbon footprint. However, manufacturing parts by adding material in layers is a painstakingly slow process when compared with volume manufacturing, particularly where polymers are concerned. Furthermore, the amount of energy required for many of the 3DP processes is quite high in relation to the number of parts, again when compared against volume manufacture. Users must therefore consider the intrinsic value of parts before a commitment to build them is made.

6 Conclusions

The use of 3D printing can definitely be considered in a frugal manner. It is an enabler for design of low-cost solutions to important problems. 3DP is a technology that can provide such solutions at many levels, including high-end problems in industries including aerospace, volume consumer products such as mobile phone covers and hearing aids, and sociably responsible products such as prosthetic hands to be used in economically-deprived regions. 3DP can also be considered in a frugal manner in terms of building low-cost and accessible machines and using recycled materials.

References

1. World Commission on Environment and Development (1987) Our common future. Oxford University Press, Oxford, p 27. ISBN 019282080X
2. Gibson I, Rosen DW, Stucker B (2009/12/30) Additive manufacturing technologies: rapid prototyping to direct digital manufacturing, 1st and 2nd edns. Springer, Berlin
3. http://home.iitk.ac.in/~abjt/assets/d/baby-carrier.pdf
4. http://forums.reprap.org/read.php?4,165263
5. http://www.wikihow.com/Make-a-Funnel-from-a-Plastic-Bottle
6. http://www.avid3dprinting.com/sls/
7. https://en.wikipedia.org/wiki/Hearing_aid
8. http://robohand.blogspot.com.au/2013/06/a-hand-for-junior.html
9. https://en.wikipedia.org/wiki/RepRap_project
10. https://www.youtube.com/watch?v=TKkXRlli-aw&feature=youtu.be

Carbon Footprint Assessment of Additive Manufacturing: Flat and Curved Layer-by-Layer Approaches

Subramanian Senthilkannan Muthu and Savalani Monica Mahesh

Abstract: This chapter describes and discusses the carbon footprint assessment of two additive manufacturing (AM) processes. The first process is the fused deposition modeling (FDM) process. The second process is the curved FDM process which is a process that has been adopted from the FDM process itself to overcome the stairstepping effects of the traditional layer-by-layer processes. The focus of this study is to explore the carbon footprint of both the technologies and its implications.

Keywords Flat layer-by-layer · Curved layer-by-layer · Carbon footprint · GHG emissions · Waste · Material consumption · Electricity · Nozzle

1 Introduction

The use of additive manufacturing (AM) for the production of final parts has been a challenging topic for researchers and developers in the last few decades. The impact of its emergence is already being felt by high value-added manufacturing industries including aerospace, military, motorsports, automotive, industrial machinery, medicine, dentistry, high-end consumer products, art, and jewelry. These industries are increasingly striving to find ways to create highly customized products at low volumes, with appropriate mechanical properties and in an affordable manner. Currently, conventional techniques still lack the ability to manufacture such products at a reasonable cost in some areas.

Although AM is gaining ground over conventional manufacturing techniques, there is still much headway to be made, especially when you consider that there are

S.S. Muthu
SGS (HK) Limited, Hong Kong, China

S.M. Mahesh (✉)
Department of Industrial Systems and Engineering,
The Hong Kong Polytechnic University, Hong Kong, China
e-mail: mfsmm@polyu.edu.hk

© Springer Science+Business Media Singapore 2016
S.S. Muthu and M.M. Savalani (eds.), *Handbook of Sustainability in Additive Manufacturing*, Environmental Footprints and Eco-design of Products and Processes, DOI 10.1007/978-981-10-0606-7_5

1000 actual products for every prototype made [1]. Although AM is now being applied to final part production, most machines are still used to fabricate prototypes. Consequently, the opportunity for much more commercial activity for AM in manufacturing applications is immense. From the manufacturers' point of view such a development will open up a whole new market; they will be able to reduce process steps for production, lower production costs, shorter lead times will be needed, and immediate cost savings may also be realized by material savings and elimination of tooling requirements. From the consumers' point of view it would be more affordable for consumers to purchase customized and high value-added products and they would also be able to create products to suit their particular needs. In addition, the exploitation of such technology would also benefit the environment, in that the customized products would be produced closer to the market which would mean reduction in transportation to and from source and destination.

Many AM techniques cannot be used to create final parts due to the precision, speed, and material property limitations of the processes involved. Researchers are now realizing new potential applications that may be vast and much less limited if forward thinking is applied to the conventional AM techniques already available. One such limitation common to all conventional AM techniques is the deposition of flat, 2D planar layers (Fig. 1).

To overcome these limitations, it would be ideal to develop new techniques that do not rely on the deposition of flat, 2D planar layers. Hence the development of novel techniques, based on current fused deposition modelling and screw extrusion systems, namely a curved additive manufacturing technique (CAMT; Fig. 2) that provides fabrication of parts with a minimum number of layers which would inherently increase the strength of parts and improve surface quality. In addition, materials that may inherently benefit from the CAMT will be developed and examined. For example, the use of short fiber-based composite materials can provide enhanced mechanical properties.

The Curved Layer Fused Deposition Modelling technique has gained a significant amount of attention as a result of its advantages such as increased flexural strength, reduction of the stairstepping effect, and the reduction in the number of layers, especially for thin shell-like structures.

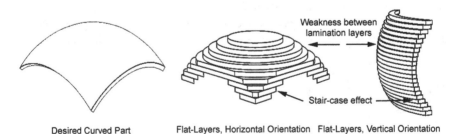

Fig. 1 The staircase effect and lamination weakness problems caused by conventional flat-layer rapid prototyping (*Source* Authors)

Fig. 2 A curved-layer part
(*Source* Authors)

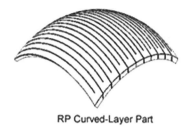

RP Curved-Layer Part

2 FDM Process

Fused deposition modeling (FDM) begins with a software process that processes an STL file (stereolithography file format), mathematically slicing and orienting the model for the build process. If required, support structures may be generated. The machine may dispense multiple materials to achieve different goals; for example, one material may be used to build up the model and another used as a soluble support structure [2], or multiple colors of the same type of thermoplastic on the same model could be used.

The model or part is produced by extruding small flattened strings of molten material to form layers as the material hardens immediately after extrusion from the nozzle (Fig. 3).

A plastic filament or metal wire is unwound from a coil and supplies material to an extrusion nozzle that can turn the flow on and off. There is typically a worm-drive that pushes the filament into the nozzle at a controlled rate. The nozzle is heated to melt the material. The thermoplastics are heated past their glass transition temperature and are then deposited by an extrusion head.

The nozzle can be moved in both horizontal and vertical directions by a numerically controlled mechanism. The nozzle follows a tool-path controlled by a computer-aided manufacturing (CAM) software package, and the part is built from the bottom up, one layer at a time. Stepper motors or servo motors are typically employed to move the extrusion head. The mechanism used is often an X–Y–Z rectilinear design, although other mechanical designs such as deltabot have been employed.

Although as a printing technology FDM is very flexible, and it is capable of dealing with small overhangs by the support from lower layers, FDM generally has some restrictions on the slope of the overhang, and cannot produce unsupported stalactites.

The carbon footprint is one of the very widely used measures to assess the environmental impact of a product or process. Due to the consequences of climate change and its effects, it is highly essential to assess the carbon footprint of different products produced by various industrial sectors. Assessment of the carbon footprint in AM is relatively new and this chapter deals with the carbon footprint assessment of two AM processes: the FDM process and the curved fused deposition (CFD).

Fig. 3 Fused deposition
modeling process (*Source*
Authors)

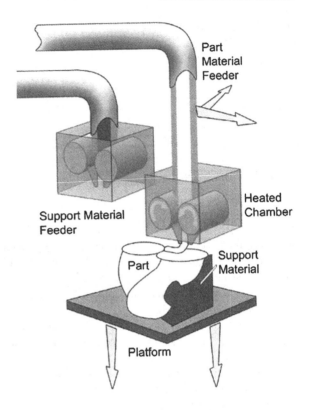

3 Experimental Design

The experimental design includes the fabrication of three-point bending samples
(shown in Fig. 4) using both the FDM process and the curved FDM process made
using the standard Up Printer filament with a diameter of 1.75 mm. To build these
parts, details of the experiment and calculations are shown in Sect. 3.1.

3.1 Experimentation and Calculation

The process flow of the experiment is illustrated in Fig. 5. Initially, a 3D model of
specific geometries in Fig. 4 was saved as an ASC‖ format STL file and then was
sliced with both the flat layer-by-layer approach and curved layer-by-layer
approach. For slicing code, layer thickness, fill gap, and printing speed, in theory
are the same for these two methods. Secondly, during the printing period, several
parameters such as voltage variation and electrical current variation of both nozzle
and printer were recorded. Finally, for the postprocessing, the sample weight and
energy cost were calculated. More details are discussed step by step as follows.

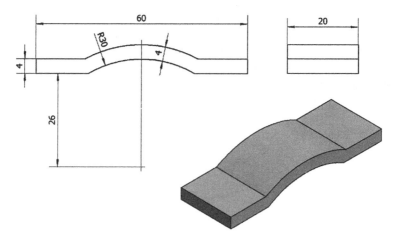

Fig. 4 Sample dimension (*Unit* mm)

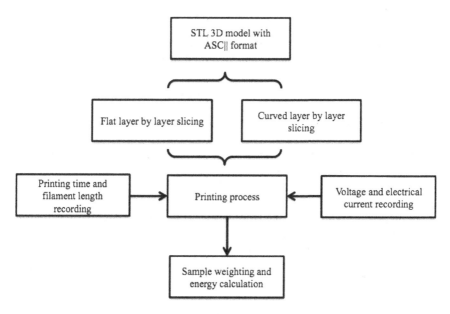

Fig. 5 Process flow

(1) Currently, there are several ways to create a 3D model. Gao, Zhang, and Ramanujan et al. [3] have summarized the most popular computational tools and development trend to generate a 3D model such as 3D optical scanning, professional 3D (integrated development environment) IDE, codesign/cocreation platform, and natural user interface (NUI)-driven shape modeling. In this experiment, Autodesk Inventor was used to generate the 3D model with dimensions shown in the materials section illustrated in Fig. 4.

(2) For coding preparation the tessellated STL 3D model with ASC∥ format was saved by Autodesk Inventor, which is readable when opening with test editor compared with binary STL. Also there are plenty of articles [2–4] published about slicing STL to generate a tool path. The data format of the STL file could be seen as follows.

facet normal x xx

outer loop

    vertex x xx   **STL information put here**

    vertex x xx

    vertex x xx

endloop

endfacet

Then the STL file is sliced using both flat and curved layer-by-layer approaches; the flat layer-by-layer slicing algorithm may be seen in Refs. [4–7] and the curved layer-by-layer slicing algorithm may be seen in Refs. [8–10]. Also in order to compare the consumed time and energy used by these two approaches, the layer thickness, printer speed and fill gap, nozzle diameter, and filament diameter are set as constant, which is shown in Table 1 [11].

(3) Regarding the printing process, in order to calculate the energy consumed by the nozzle and printer, the loop current and loop voltage of both the nozzle and printer circuit should be read. To achieve this, we connect the multimeter in serial within the printer circuit. In this way, the loop current is shown in units of amperes. Additionally, the loop voltage of the printer circuit is determined by the power adaptor. The same principle is used in the nozzle heating circuit. Also, to calculate the printing time for each sample, a stopwatch is used.

(4) In postprocessing, after we get all the specimens, the weight of both samples, the curved structure of the 3D model, and the supports should be calculated. However, there are two approaches to achieve this. One way is to weigh the sample and corresponding support, respectively, and another way is to weigh the whole specimen with weight M1, and after removing the support, weigh

Table 1 Build parameters

Approaches	Fill gap (mm)	Filament dia. (mm)	Nozzle dia. (mm)	Layer thickness (mm)
Planar layer FDM	0.7	1.75	1	1
CLFDM	0.7	1.75	1	1

Table 2 Weight of samples and corresponding supports

Approach	Sample weight (g)	Material waste (support)	Total material consumption (g)
Flat layer-by-layer approach	4.82	0.665	6.0335
	4.92	0.662	6.1402
	4.9	0.663	6.1193
	4.91	0.665	6.1325
	4.88	0.665	6.0995
Curved layer-by-layer approach	5.47	0.665	6.7485
	5.47	0.664	6.7474
	5.5	0.665	6.7815
	5.48	0.665	6.7595
	5.44	0.663	6.7133

the sample again with weight M2, hence the weight of the sample is M1 − M2. Because during the process of removing the support, the support is broken and some fragments are hard to be collected, we therefore use the second approach to obtain the weight of samples and support. The data of both the flat layer and curved-layer printed samples and corresponding supports are listed in Table 2.

Calculation:

Based on physical principle, energy E generated in the circuit could be expressed as

$$E = UIt$$

where U is the voltage supplied to the circuit, I is the loop current, and t is the continuing time.

Thus, the energy consumed by nozzle E_n is

$$E_n = U_n I_n t_n$$

where U_n is the voltage supply to the nozzle, I_n is the electrical current read by the multimeter, and t_n is the printing time read from the stop watch.

Additionally, the energy consumed by printer E_p is

$$E_p = U_p I_p t_p$$

where U_p is the voltage supply to the printer, I_p is the electrical current read by the multimeter, and t_p is the printing time read from the stop watch.

Hence the total energy E_t consumed by fabricating one piece of the sample is

$$E_t = E_n + E_p$$

4 Carbon Footprint Assessment

4.1 Goal of the Study

The goal of this product carbon footprint (PCF) study is to calculate the potential contribution of the chosen product produced by two different AM techniques to global warming expressed as grams or kilograms of CO_2 eq., by quantifying all significant GHG emissions and removals over the product's life cycle (only "cradle to gate" stages are considered in this study; i.e., the scope of this study is pertaining to cradle to gate only). This report follows the standard of ISO/TS 14067: 2013 "Greenhouse gases—Carbon footprint of products—Requirements and guidelines for quantification and communication" [12]. The results, data, methods, assumptions, and life-cycle interpretations are presented in sufficient detail to allow the reader to comprehend the complexities and trade-offs inherent in the PCF study. The PCF value is measured by quantifying the GHG emissions for the entire product life cycle expressed in carbon dioxide equivalents value (kg CO_2 eq.), as per ISO/TS 14067.

$$CF_{total} = CF_{RM} + CF_M + CF_D + CF_U + CF_{EOL} \tag{1}$$

where CFRM is the carbon footprint of the raw material stage (kg CO_2 eq.); CFM is the carbon footprint of the manufacturing stage (kg CO_2 eq.); CFD is the carbon footprint of the distribution stage (kg CO_2 eq.); CFU is the carbon footprint of the use stage (kg CO_2 eq.); and CFEOL is the carbon footprint of the end-of-life stage (kg CO_2 eq.).

5 Scope

5.1 Product System and Its Function(s)

The product selected for this study is an identical product produced by two different AM techniques, namely flat and curved layer-by-layer approaches.

5.2 Functional Unit

The functional unit is defined as one piece of the sample produced at a laboratory scale, made out of ABS-acrylonitrile butadiene styrene (3-point bending sample) produced by two different AM techniques (average sample weight of flat layer-by-layer approach is 4.886 g and curved layer-by-layer is 5.472 g). A small difference in the weight of identical samples is attributed to the virtue of the manufacturing techniques (way of building layers). The reference flow is defined as one piece of the 3-point bending sample produced by two different AM techniques.

5.3   Data for the Study

This is a simple screening PCF study covering the cradle to gate stages of the products produced by the two techniques mentioned above. Primary data in terms of weight of the sample, amount of energy needed to produce a product, time to build a product, material waste, material consumption, and efficiency were collected (from the laboratory scale as explained above).

Emission factors for the calculation were obtained from the secondary databases, explained later in detail.

5.3.1   Characterization of Data Quality

The best quality data were used to reduce bias and uncertainty as far as practically possible. Primary and secondary data were selected to meet the goal and the scope of the PCF study.

5.4   Cut-Off Criteria and Cut-Off

Apart from the primary data elements mentioned in Sect. 5.3.1, the rest of the data such as the packaging of the ABS, transportation of the ABS, and so on were not included.

5.5   Allocation Procedures

The allocation procedure is not applicable for this screening study.

5.6   Geographical and Time Boundary of Data

The ABS samples are produced in the HK Polyu's lab and hence the emission factor for electricity is taken to represent the emission situation of HK.

5.7 System Boundary

The system boundary for this study covers the cradle to gate of the ABS sample products' life-cycle stages, from raw material extraction to manufacturing. As discussed in Sect. 5.4, apart from the raw material and manufacturing energy, the rest of the elements associated with this study are considered to be out of the scope. All unit processes included in this study directly contributed to the end result (GHG emissions) from cradle to gate life-cycle stages.

5.8 Assumptions

There are no specific assumptions in this study as this study is a very simple and straightforward one except for the experimental assumptions.

5.9 Treatment of Electricity

The emission factor of electricity is obtained from the carbon emission intensity figures of 2014 from CLP (China Light and Power), where the products in question are being manufactured. Emissions of electricity produced by CLP obtained from the latest annual report of CLP are 0.640 kg CO_2 eq./kWh [13].

6 Methodology

According to the data collected from the laboratory conditions, calculation of the cradle to gate assessment of the product's carbon footprint is based on the LCA approach, which includes the raw material stage and the manufacturing stage The analysis is based on the concept of life-cycle assessment (LCA) and ISO/TS 14067: 2013 [12], and allows computation of the total amount of carbon emission within the selected system boundary. This PCF study, according to ISO/TS 14067: 2013, includes four phases of LCA: that is, goal and scope definition, life-cycle inventory (LCI), life-cycle impact assessment (LCIA), and life-cycle interpretation (LCI). Figure 6 shows the framework of LCA from ISO 14040 [14].

The unit processes comprising the product system are grouped into selected life-cycle stages in this study, that is, raw material acquisition and manufacturing. GHG emissions and removals from the product's life cycle are assigned to the life-cycle stage in which the GHG emissions and removals occur. Partial PCFs may

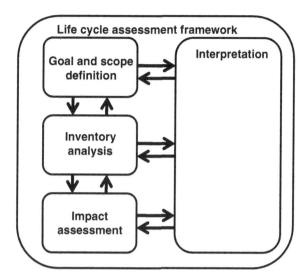

Fig. 6 Framework of life-cycle assessment [14]

be added together to quantify the PCF, provided the same methodology is used and there are no gaps and overlaps. The PCF is calculated and presented as a carbon dioxide equivalent (kg CO_2 eq.) using the relevant 100-year global warming potential (GWP 100). Findings in the impact assessment and the sensitivity and contribution analysis, performed as part of the interpretation, help identify the most relevant and contributing ("key") processes and basic flows of the system.

$$Carbon\ Emission = Activity\ Data \times Emission\ Factor \times Global\ Warming\ Potential$$

(2)

The "Global Warming Potential" should be excluded, where the unit of the "Emission Factor" was "kg CO_2 eq./unit".

7 Life-Cycle Inventory

Life Cycle inventory collected from the experimentation part is tabulated in Tables 3 and 4. As far as emission factors are considered, they were obtained from credible databases. Database selection was based on credibility and geographical, technological, and time-related considerations. The emission factor for electricity consumption was obtained from the carbon emission intensity figures of 2014 from CLP (China Light and Power), where the products in question are being manufactured as explained in Sect. 5.9. The emission factor for ABS was obtained from Ecoinvent version 2.2.

Table 3 Life cycle inventory

Approach	Material	Sample weight (g)	Build time (min)	Power (Kwh/piece)	Material waste (support)	Material consumption (g)	Efficiency (%)	Total material consumption (g)
Flat layer-by-layer approach	ABS	4.82	8:40	0.00638	0.665	5.485	90	6.0335
	ABS	4.92	8:40	0.0064	0.662	5.582	90	6.1402
	ABS	4.9	8:40	0.0064	0.663	5.563	90	6.1193
	ABS	4.91	8:40	0.0064	0.665	5.575	90	6.1325
	ABS	4.88	8:40	0.0064	0.665	5.545	90	6.0995
Curved layer-by-layer approach	ABS	5.47	9:58	0.0074	0.665	6.135	90	6.7485
	ABS	5.47	9:58	0.0074	0.664	6.134	90	6.7474
	ABS	5.5	9:58	0.0074	0.665	6.165	90	6.7815
	ABS	5.48	9:58	0.0074	0.665	6.145	90	6.7595
	ABS	5.44	9:58	0.0074	0.663	6.103	90	6.7133

Table 4 Life cycle inventory: average results

	Flat layer-by-layer approach	Curved layer-by-layer approach
Material consumption in g/piece	6.105 g	6.750 g
Electricity in Kwh/piece	0.0064	0.0074

8 Product Carbon Footprint Calculation and Interpretation

The results of PCF calculation are presented in Fig. 7. From the PCF calculation results, it is very clear that the flat layer-by-layer approach is emitting lower GHG emissions compared to the curved technique. It is also noted that there is a small weight difference between the samples owing to the virtue of the layer building technique; again, this is also one of the criteria of the carbon footprint which shows the flat method outweighs the curved technique in terms of PCF. Electricity (energy required) is a notable parameter, which is less for the flat method compared to its rival, which has also resulted in the lower GHG emissions.

It is a simple study, again, a screening study involving two important parameters, material consumption and electricty. A further detailed LCA and a PCF study are recommended for further research to discover more insights on the environmental impacts of these two AM techniques.

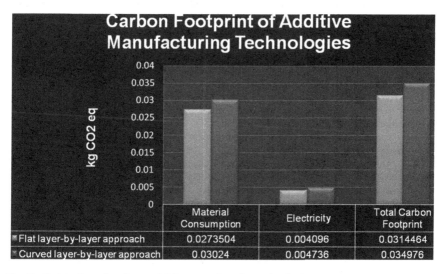

Fig. 7 Carbon footprint of two additive manufacturing technologies

9 Conclusion

The study has shown that the FDM process is environmentally more friendly than the curved FDM process in terms of material usage, wastage, and build time of parts. However, the authors have studied the mechanical advantages of using the curved FDM process in comparison to the FDM process. As a result the authors recommend a study of the environmental impact in more detail.

Acknowledgements We would like to thank the Hong Kong Polytechnic University for supporting this project under the code G-YM77.

References

1. Wohler's Report (2010) An in-depth global study on the advances in additive manufacturing technologies and applications. ISBN:0-9754429-6-1
2. Wen PZ, Huang WM, Wu C-K (2008) Modified fast algorithm for STL file slicing. J Comput Appl 28:1766–1768
3. Gao W, et al (2015) The status, challenges, and future of additive manufacturing in engineering. Comput Aided Des 69:65–89
4. Wuyi ZBWSC (2004) Algorithm for rapid slicing STL model. J Beijing Univ Aeronaut Astronaut 4:011
5. Brown AC, De Beer D (2013) Development of a stereolithography (STL) slicing and G-code generation algorithm for an entry level 3-D printer. In: Africon, 2013. IEEE
6. Choi SH, Kwok FKT (1999) A memory efficient slicing algorithm for large STL files. In: Proceedings of solid freeform fabrication symposium
7. Tata K, et al (1998) Efficient slicing for layered manufacturing. Rapid Prototyp J 4(4):151–167
8. Chakraborty D, Reddy BA, Choudhury AR (2008) Extruder path generation for curved layer fused deposition modeling. Comput Aided Des 40(2):235–243
9. Klosterman DA, et al (1999) Development of a curved layer LOM process for monolithic ceramics and ceramic matrix composites. Rapid Prototyp J 5(2):61–71
10. Singamneni S, et al (2012) Modeling and evaluation of curved layer fused deposition. J Mater Process Technol 212(1):27–35
11. Guan HW, et al (2015) Influence of fill gap on flexural strength of parts fabricated by curved layer fused deposition modeling. Procedia Technol 20:243–248
12. International Organisation for Standardization (ISO) (2013) ISO/TS 14067:2013 greenhouse gases—carbon footprint of products—requirements and guidelines for quantification and communication
13. CLP Annual Report (2014) https://www.clpgroup.com/en/Sustainability-site/Report%20Archive%20%20Year%20Document/SR_In_Essense_2014_en.pdf
14. Interational Organisation for Standardization (ISO) (2006) ISO 14040: 2006 environmental management—life cycle assessment—principles and framework

Printed in the United States
By Bookmasters